Cambridge Elements

Elements in Organizational Response to Climate Change
edited by
Aseem Prakash
University of Washington
Jennifer Hadden
University of Maryland
David Konisky
Indiana University
Matthew Potoski
UC Santa Barbara

GREENING THE INTERNATIONAL MONETARY FUND

Alexandros Kentikelenis
Bocconi University

Thomas Stubbs
Royal Holloway, University of London

CAMBRIDGE
UNIVERSITY PRESS

Shaftesbury Road, Cambridge CB2 8EA, United Kingdom

One Liberty Plaza, 20th Floor, New York, NY 10006, USA

477 Williamstown Road, Port Melbourne, VIC 3207, Australia

314–321, 3rd Floor, Plot 3, Splendor Forum, Jasola District Centre, New Delhi – 110025, India

103 Penang Road, #05–06/07, Visioncrest Commercial, Singapore 238467

Cambridge University Press is part of Cambridge University Press & Assessment, a department of the University of Cambridge.

We share the University's mission to contribute to society through the pursuit of education, learning and research at the highest international levels of excellence.

www.cambridge.org
Information on this title: www.cambridge.org/9781009697538

DOI: 10.1017/9781009697491

© Alexandros Kentikelenis and Thomas Stubbs 2025

This publication is in copyright. Subject to statutory exception and to the provisions of relevant collective licensing agreements, with the exception of the Creative Commons version the link for which is provided below, no reproduction of any part may take place without the written permission of Cambridge University Press & Assessment.

An online version of this work is published at doi.org/10.1017/9781009697491 under a Creative Commons Open Access license CC-BY-NC-ND 4.0 which permits re-use, distribution and reproduction in any medium for non-commercial purposes providing appropriate credit to the original work is given. You may not distribute derivative works without permission. To view a copy of this license, visit https://creativecommons.org/licenses/by-nc-nd/4.0

When citing this work, please include a reference to the DOI 10.1017/9781009697491

First published 2025

A catalogue record for this publication is available from the British Library

ISBN 978-1-009-69753-8 Hardback
ISBN 978-1-009-69752-1 Paperback
ISSN 2753-9342 (online)
ISSN 2753-9334 (print)

Cambridge University Press & Assessment has no responsibility for the persistence or accuracy of URLs for external or third-party internet websites referred to in this publication and does not guarantee that any content on such websites is, or will remain, accurate or appropriate.

For EU product safety concerns, contact us at Calle de José Abascal, 56, 1°, 28003 Madrid, Spain, or email eugpsr@cambridge.org

Greening the International Monetary Fund

Elements in Organizational Response to Climate Change

DOI: 10.1017/9781009697491
First published online: October 2025

Alexandros Kentikelenis
Bocconi University

Thomas Stubbs
Royal Holloway, University of London

Author for correspondence: Thomas Stubbs,
thomas.stubbs@rhul.ac.uk

Abstract: The International Monetary Fund (IMF) has emerged as a key player in climate policy. The organization introduced its Climate Strategy in 2021 and established the Resilience and Sustainability Facility in 2022 to provide financial support to countries facing adaptation and mitigation challenges. The IMF's closer engagement with the economic dimensions of climate change holds the promise of helping countries preempt large-scale economic dislocations from climate risks. But how much progress has the IMF made in supporting the green transition? What is the policy track record of the IMF's climate loans? How do regular IMF loans and mandated reforms encompass climate considerations? How have the IMF's economic surveillance activities considered climate risks? Based on new evidence, the findings in this Element point to the multifaceted, and at times contradictory, ways green transition objectives have become embedded within IMF activities. This title is also available as Open Access on Cambridge Core.

Keywords: green transition, International Monetary Fund, global economic governance, climate policy, climate finance

© Alexandros Kentikelenis and Thomas Stubbs 2025

ISBNs: 9781009697538 (HB), 9781009697521 (PB), 9781009697491 (OC)
ISSNs: 2753-9342 (online), 2753-9334 (print)

Contents

1 The IMF and the Green Transition — 1

2 Organizational Innovation: Introducing the Resilience and Sustainability Facility — 12

3 Organizational Inertia: The IMF's Standard Lending Agreements — 21

4 Organizational Experimentation: Economic Surveillance and Climate Advice — 36

5 Conclusions: Real but Limited Organizational Change — 49

Appendix — 56

References — 63

1 The IMF and the Green Transition

1.1 An Evolving Engagement with Climate Issues

When the world's senior economic policymakers met in Washington DC in April 2022, they were ready to agree on bold action at the intersection of climate change and international financial management. The two years of economic turmoil after the onset of the deadly COVID-19 pandemic gave rise to a sense of urgency for embedding climate considerations into economic recovery efforts. The tone was already set in early policy initiatives by the Biden administration in the United States to 'build back better' and by the European Union's promotion of a 'European Green Deal.' Among such policy optimism, participants in the Spring Meetings of the International Monetary Fund (IMF) and the World Bank celebrated the creation of a new lending instrument to support countries in implementing climate-related economic policies that would help them address the structural challenge of climate change.

But this climate orientation did not immediately or universally appear as an obvious activity for the IMF – the central actor in global economic governance – to be involved in. The IMF's mandate is to underpin global financial stability in two ways: by providing financial assistance to countries' unsustainable balance of payments positions (i.e., those who struggle to meet their external financial obligations), and by conducting periodic monitoring of its members' economic policies to pre-empt macroeconomic problems from emerging. Neither of these roles directly relates to considerations about the environment, and the path toward implementing organizational reforms that foreground climate considerations had been long. As early as 2015, then-Managing Director Christine Lagarde (2015) explained that climate issues are 'macro-critical,' impacting the economy as a whole and, therefore, within the Fund's remit. This led to a flurry of activity – primarily analytical – to spell out economic policy measures to underpin the green transition (Clift 2024). By 2023, the IMF's current Managing Director Kristalina Georgieva could confidently reiterate that "climate risks affect macroeconomic and financial stability" and inform the world that "we are a financial institution, so we put money where our mouth is" (World Bank 2023b). These issues were now dominating the policy agenda, and the IMF was establishing itself at the forefront of debates on the intersection of economic and environmental policies.

To transform high-level pronouncements into concrete organizational action, the IMF has made several strides. At the broadest level, the organization introduced its Climate Strategy in 2021, where it committed to scale up its engagement with macro-critical climate issues, primarily by expanding its analytical capabilities and incorporating climate considerations into its regular multilateral and bilateral economic surveillance (IMF 2021d). In the same year, the Comprehensive Surveillance Review committed to expand analyses on climate issues in line with the goals set out in the Paris Agreement, as well as to recruit additional staff with climate-relevant economic expertise (IMF 2021b). And in 2022, the IMF developed a new lending facility – the Resilience and Sustainability Facility (RSF) – to provide financial support primarily to climate-vulnerable countries that face steep adaptation and mitigation challenges.

These are only the direct ways in which the IMF has become a central player in climate debates within the global governance architecture. To be sure, as a financier of climate adaptation and mitigation strategies, the role of the IMF is relatively limited compared to that of development banks (Clifton, Fuentes, and Howarth 2021; Humphrey 2022; Kentikelenis and Babb 2022; Mertens and Thiemann 2017, 2018; Naqvi, Henow, and Chang 2018). However, the centrality of the IMF arises due to its ability to also *indirectly* influence the direction of climate policy. On the one hand, this is linked to the IMF's role in shaping economic policy debates at the national level (Nelson 2014, 2017; Reinsberg et al. 2019). Rather than providing financing for projects – like the development of solar power infrastructures or the restoration of wetlands – that may be expensive but are ultimately clearly demarcated, the IMF's ambition is to help steer macroeconomic policy. This means that its reach can be more thorough and longer lasting than the amount of money it offers would suggest. On the other hand, the IMF's epistemic authority sets the tone of debates and marks the range of 'legitimate' economic policies for combatting climate change (Broome and Seabrooke 2015).

These developments in the IMF's role are important for global efforts to combat climate change, but they also beg important questions concerning the evolution of international institutions to incorporate environmental considerations into their organizational practices. What is the nature of the IMF's involvement with climate issues? Has the organization meaningfully integrated climate considerations into all aspects of its activities? And to what extent has the new modus operandi supplanted previous practices? We tackle these questions using quantitative and case study evidence on the IMF's activities over the 2020–2024 period.

1.2 The IMF at the Intersection of Macroeconomics and Climate

In the aftermath of the COVID-19 pandemic, developing countries have faced not only domestic economic and social fallout but also a highly destabilizing global economic environment. Pandemic-related slowdowns were followed by fiscal crises, exacerbated by surging energy and food prices. Interest rate hikes by central banks in high-income countries further intensified these pressures, driving up debt service costs (Chowdhury and Sundaram 2023; Fischer and Storm 2023). By October 2024, thirty-five low- and lower-middle-income countries were in debt distress or at high risk of it, and many more have brewing debt concerns (IMF 2024g). Against the backdrop of intense fiscal and debt pressures and mounting social dislocations, many developing countries have turned to the IMF for support, whether to provide financial assistance (Stubbs et al. 2021) or to offer policy advice as part of its regular economic monitoring function (Breen and Doak 2023; Gallagher et al. 2021).

These activities of the IMF are both influential and controversial. For its part, lending to countries facing economic problems has been the subject of recurring debate due to the IMF's powerful tool of 'conditionality.' That is, in exchange for financial assistance at low interest rates and long maturities, countries that turn to the organization must commit to implement extensive policy conditions that are updated and expanded over the life course of the lending program (generally one to four years and reviewed quarterly or biannually). These conditions invariably cover a range of fiscal and financial targets, like targets for the budget deficit, the nonaccrual of debt arrears, or the level of foreign reserves. But they also include so-called 'structural' reforms: measures that seek to directly alter the economic structures of the borrowing country, like the deregulation of certain economic activities, the privatization of state-owned enterprises, or the liberalization of labor markets (Kentikelenis and Stubbs 2023). Thus, conditionality is the coercive toolkit of the institution and gives this international organization unprecedented power over the domestic policy environments of its borrowers (Simmons, Dobbin, and Garrett 2008).

The IMF's economic surveillance activities are also highly consequential. These are mandated by the organization's founding treaty – the Articles of Agreement (IMF 2020a) – and performed annually for large economies or biennially for most other countries. Teams of IMF staff visit countries to collect and analyze data and conduct consultations with domestic policymakers (primarily from ministries of finance and central banks). Subsequently, they publish their candid assessment of a country's policy environment and provide recommendations on what reforms should be implemented. These are known as 'Article IV reports,' named after the corresponding clause in the IMF's founding treaty. The policy recommendations included in these reports do not have the coercive character of IMF conditionality, but they nonetheless have an important role in shaping both domestic policymakers' priorities and international markets' evaluation of country investment potential and creditworthiness (Breen and Doak 2023).

1.2.1 Debt, Debt Sustainability, and Climate Risks

The IMF's growing engagement with climate issues takes place against a backdrop of deepening debt problems in developing countries. This is important because of the many transmission channels between climate risk and debt. First, climate change has a range of follow-on implications for fiscal sustainability, financial sector stability, and overall macroeconomic health, which in turn affect debt sustainability (Ramos et al. 2022a; Volz and Ahmed 2020a). Second, countries with high degrees of climate vulnerability also have to pay risk premiums on their sovereign debt, which inevitably increases their overall debt burden (Kling et al. 2021). Indicatively, V20 countries – a group of climate-vulnerable nations – faced average real interest rates of 9% in the mid-2010s, compared to 6% for G20 members, the group of twenty richest economies in the world (Volz and Ahmed 2020a). Third, high levels of debt service hamper the introduction of green transition policies, as public funds get directed to paying off external debt rather than financing climate change adaptation or mitigation projects (Ramos et al. 2022c). Debt problems also prompt countries into development trajectories that are antithetical to green transition objectives. Civil society has pointed to "a debt – fossil fuel production trap, whereby countries rely on fossil fuel revenues to repay debt, and anticipated revenues from fossil fuels are often overinflated and require huge investments to reach expected returns, leading to further debt, eroding long-term development prospects, and causing devastating environmental and human harms" (Woolfenden 2023a).

In other words, there is a deep and reciprocal relationship between climate change and debt, whereby climate risks contribute to debt problems, and debt problems make dealing with climate risks even more difficult in resource-constrained environments (ActionAid International 2023). A closer look at debt service statistics of climate-vulnerable countries reveals the scale of the problem. Figure 1 shows the average of external debt service (as a share of GDP and excluding debt to the IMF) over the decade preceding the pandemic (circle) and compares it to its level in 2024 (diamond). While countries at all levels of climate vulnerability have registered increases in their external debt service, this phenomenon is most pronounced for those that are highly vulnerable. The average highly climate vulnerable country directed 1.2% of its GDP to debt service in the 2010s but now spends 3% for the same function. This inevitably raises important questions around fiscal priorities and suggests that the approach of policymakers to debt problems also matters for their ability to pursue green transition objectives. In a time of ballooning debt service, longer-term investments – like those necessary for meeting climate goals – could be the first to be scaled back.

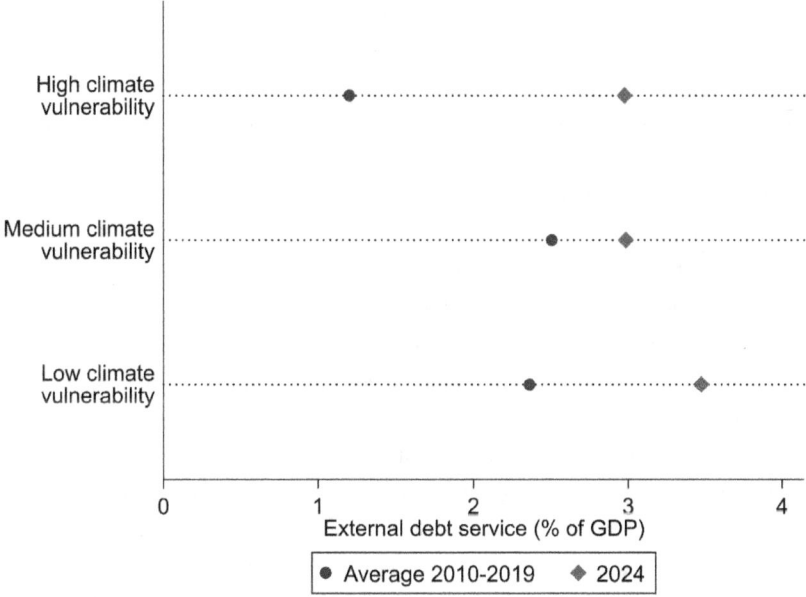

Figure 1 External debt service, disaggregated by countries' climate vulnerability

Source: Authors, using data from the IMF (2023j) and the University of Notre Dame (2023). External debt service includes public and publicly guaranteed debt but excludes IMF charges and repurchases.

Helping countries deal with such debt problems is a core function of the IMF. To this end, the organization employs Debt Sustainability Analyses (DSAs) to shape its activities in this area. The methodology for these analyses depends on the income classification of the evaluated country: one methodology applies to low-income countries, while another to the remaining so-called 'market access' countries. These evaluations are important because if the IMF considers a country's debt unsustainable, it must stop lending and propose debt restructuring. Yet, such negative assessments are uncommon (Rehbein 2023). A pre-pandemic review at the IMF showed that highly debt-vulnerable countries generally did not pursue restructurings, even though the latter were associated with a greater probability of the IMF's lending being successful (IMF 2019a).

Given the macroeconomic implications of the climate crisis, the IMF has started incorporating climate issues within DSAs. For example, the recently updated framework for market access countries ('Sovereign Risk and Debt Sustainability Framework') has introduced a two-part climate change module (IMF 2022m). The first component examines the impact of adaptation investments, and the other focuses on mitigation policies. Based on the scenarios pursued in these modules, the outcome is projections for debt-to-GDP and gross financing needs-to-GDP ratios, which – in turn – illustrate debt-related risks from green transition policies. These projections also shape the kinds of reforms that are advocated within IMF lending programs, thus foregrounding the centrality of this debt toolkit in the IMF's engagement with climate issues.

The analyses and assumptions included in DSAs remain the subject of sustained debate. In relation to climate issues, a recent analysis pointed out the limited consideration of different physical or transition risk scenarios in the IMF's modelling efforts (Maldonado and Gallagher 2022). This matters because once different risks are considered, and the introduction of baseline or greener transition policies modeled, then the findings on government financing needs and public debt can vary considerably. Beyond the inclusion of climate issues, debt sustainability analyses are not an exact science but an exercise that entails a high degree of subjective judgement by IMF staff (Hagan 2023), and also one that is heavily politicized due to its momentous implications for opening up a path for debt restructuring (Wheatley 2023).

The climate module within DSAs is not the only set of analytical modules that the IMF has put forth. In recent years, the IMF introduced the Debt–Investment–Growth and Natural Disasters (DIGNAD) toolkit, which examines the economic implications of physical risks. In relation to debt, this toolkit enables an assessment of debt vulnerabilities in scenarios of both the aftermath of a natural disaster and the introduction of "ex-ante policies, such as building adaptation infrastructure, increasing fiscal buffers, or improving public investment efficiency" (IMF 2024c).

The build-up of growing debt vulnerabilities combined with inadequate analytical toolkits pose major challenges for crafting development policies and pre-empting debt crises. Indeed, the forthcoming review of the DSA framework for low-income countries is expected to grapple with how to expand the remit of analysis and make this instrument fit for purpose, in light of the economic challenges posed by climate change (Fresnillo 2024). In following this path, however, the IMF is not alone. In recent years, the World Bank has assumed an ever-greater role in analyzing how countries can integrate the fight against climate change with the pursuit of developmental objectives. In particular, in 2021 the Bank launched its own climate-related evaluations through the Country Climate and Development Reports (CCDR).[1] This matters because

---

[1] The relationship between the IMF and the World Bank in performing climate policy evaluations has been at times collaborative and occasionally competitive. When the IMF and the World Bank both sought to participate in global efforts to combat climate change in the mid-2010s, one of their key innovations was to introduce a joint analytical toolkit for countries: the Climate Change

many of the IMF's climate-related measures are drawn directly from these reports.

1.2.2 Fiscal Consolidation, Market-Based Approaches to the Green Transition, and Their Limits

There is a fundamental tension in IMF lending: the organization's programs demand that borrowers introduce far-ranging fiscal consolidation measures – commonly known as 'austerity' – with the hope of improving their macroeconomic positions, yet these are the policies that constrain countries' ability to introduce green transition policies. This occurs through several pathways. First, and most obviously, austerity measures limit available public financing for climate change adaptation and mitigation. Public funds to support a range of green policies – like phasing out coal, incentivizing a switch to cleaner energy sources, scaling up production of renewable energy sources, or providing green subsidies – are likely unavailable for countries introducing extensive budget cuts, and that need to decide between cuts to politically sensitive areas of public spending (e.g., education and health) versus policies that only have a pay-off in the medium- to long-run.

Second, austerity measures are also likely to stunt the introduction of any green industrial policies. Such policies include incentives for research and development and tax relief for green investments by the private sector and require active state policies to appropriately and effectively channel green investments. Absent such investments, countries run the risk of relying primarily on private finance to pursue a green transformation, which may not find it sufficiently profitable to do so or on the financing available through multilateral institutions and bilateral donors, which may come with strings attached.

Policy Assessment (CCPA). Between 2017 and 2020, six such evaluations were conducted for small island developing states on a pilot basis. Given high demand for such a product, the IMF planned to continue and scale up CCPAs to expand the coverage of countries beyond small states. Despite early optimism for the prospects of such cooperation, World Bank leadership decided to pause their engagement with the IMF on the CCPA. This was a source of concern within the IMF, as staffers considered this a duplication of work that the IMF was already performing and a step back for enhancing coordination between the two institutions (Kentikelenis, Stubbs, and Reinsberg 2022). The World Bank launched its own Country Climate and Development Reports (CCDR) in 2021. In response, the IMF developed the Climate Macroeconomic Assessment Program (CMAP) with a focus on its comparative advantage in macro-fiscal analysis, public financial management, and tax policies. Two pilots were conducted between 2021 and 2022. After the review of the pilots, the IMF concluded that the Fund would provide streamlined CMAPs in exceptional cases while expanding targeted capacity development, in order to avoid duplication of the work. The IMF now draws on the World Bank's CCDRs to inform the analyses and policies included in recent loan agreements, while using more targeted toolkits such as the Climate Public Investment Management Assessment (IMF 2024a) and the Climate Policy Diagnostics.

Finally, austerity also has more insidious effects on the likelihood of the green transition. A long-established strand of academic scholarship has shown that the economic performance of countries implementing IMF programs tends to suffer (Dreher 2006). There are many reasons for this, and they go beyond the observation that countries that resort to IMF programs are generally already going through crisis before resorting to the organization's lending. Ineffective or overambitious IMF-designed policies, inappropriate pacing of reforms, and poor analytical frameworks (e.g., in estimating the fiscal multipliers that inform the design of austerity measures) can all deepen and prolong pre-existing economic troubles (Blanchard and Leigh 2014; Blyth 2013; Kentikelenis and Stubbs 2023). Depressed economic activity has climate-damaging follow-on implications. Households find it unaffordable to shift to greener consumption patterns, with low-income ones being particularly hard-pressed. Private firms struggle to invest in green technologies and adaptation measures, thereby increasing their vulnerability to climate shocks and forestalling reductions in their emissions. And the public sector is further starved of revenue sources.

Beyond fiscal policy, IMF programs also target a broader set of policies that seek to alter countries' economic policy profiles and the relative role of the state versus the market in shaping development policies. For example, the IMF has a track record of promoting carbon pricing as a key policy to underpin climate change adaptation and mitigation. Such policies can indeed be powerful tools in the fight against climate change, but they can only be one part of a broader approach toward the green transition. As the IMF's own research has shown, implementing green transition policies in developing countries also entails increasing low-carbon manufacturing capabilities domestically (Prasad et al. 2022) and providing public support for research and development (Bettarelli et al. 2023).

1.2.3 Subsidies and Social Spending: Tools for a Just Green Transition?

In the IMF's view, removing energy subsidies forms a key plank of the proposed path toward the green transition (de Mooij, Keen, and Parry 2012). This is because such a reduction purportedly has doubly positive effects: it reduces the fiscal deficit by removing a drain on the public budget, and it reduces energy consumption by ensuring that prices reflect the true cost of carbon. While the economic logic of subsidy removal might be compelling and these policies are commonly included in IMF loan conditionality, such measures have clear social and political implications. Indeed, recent political turmoil in Ecuador and Sri Lanka is linked directly to government decisions to remove subsidies (International Crisis Group 2024; Ioanes 2023).

Discussion of subsidies needs to distinguish between two types of intended recipients. First, subsidies can be directed to firms to support their productive activities and protect them from high energy prices. As a sort of industrial policy, this approach can have beneficial effects on economic activity, even if it does not expose firms to the true price of carbon. But there is no guarantee that this in-kind support to producers will even be reinvested into productive, development-enhancing activities – they may just end up accumulating profits. Consequently, policymakers face two simultaneous challenges: how to phase out energy subsidies to producers while also ensuring that the additional costs are not passed on to consumers (here, taxation of past hyper-profits in the form of windfall taxes can help raise public revenues to compensate consumers for higher prices); and how to ensure that new industrial policies are 'green,' thus not hampering domestic or global efforts to combat climate change.

Second, energy subsidies can be directed to consumers, which is a de facto social policy in many developing countries where other forms of redistribution and social provision are severely constrained (Hosan et al. 2023). A common problem is that these subsidies are often not targeted, thus favoring those with higher incomes in absolute terms and contributing to widened inequalities. Even so, individuals with lower incomes still benefit substantially from these policies, as fuel expenditure tends to be a high proportion of their overall spending and these subsidies may be the only type of support they receive from the government (IMF 2020c). Thus, even though fuel subsidies do not expose individuals to the true cost of carbon, they help maintain social peace and provide substantial support to low-income individuals. However, while energy subsidy removal can be an important tool to combat excessive fuel consumption, the IMF's approach to fuel subsidies is their wholesale elimination and replacement by targeted social assistance measures.

1.3 The Argument

Thus far, we have covered the interconnections between climate change and the economy and emphasized the novel role the IMF has carved out for itself in supporting its members' green transition objectives. What is at stake in this organizational transformation is clear: the attainment of global targets on climate change adaptation and mitigation. So, to what extent is the IMF now 'green'? Or, in more analytical terms, have its public commitments to facilitating the green transition and its analytical work in that direction also translated into concrete and coherent organizational policies and practices?

The short answer to these questions is that 'it depends.' The IMF is not a monolithic organization, and neither is the incorporation of a climate agenda in its operations. Drawing on extensive new data collection and in-depth case studies of country experiences, we demonstrate that climate change considerations have unequally infused different aspects of the IMF's work. This is because the spread of such practices within the organization has occurred at different speeds: some new areas of IMF operations have proven amenable to fast-tracking the inclusion of policies that staff consider climate-friendly, whereas more traditional aspects of its operations – most notably, the bulk of its lending – has not been meaningfully updated to take into account climate goals.

We schematize our findings into three distinct processes within the organization. First, we find clear evidence of *organizational innovation* vis-à-vis the embedding of climate considerations in the IMF's new climate-oriented lending instrument, the Resilience and Sustainability Facility. While lending under this facility accounts for a small fraction of total IMF commitments thus far, it has proven attractive among potential borrowers and is a major initiative – relying on fresh financial resources – to support climate-friendly structural change in countries' policy environments. Agreements under this new lending facility focus on fiscal policy, sectoral reforms, and private finance mobilization for climate objectives. Even though such measures have the potential to increase policy space for climate action, they overwhelmingly favor the introduction of market-oriented reforms and direct countries away from state-led models of green transformation, at the same time that such policies are actively pursued by large economies around the world.

Second, the IMF's traditional loan programs – representing the bulk of the organization's lending operations – demonstrate *organizational inertia*. That is, while there are frequent references to climate considerations in the loan documentation, the IMF tends to prioritize fiscal consolidation, deregulation, and market-oriented reforms, often at odds with the investments and policy shifts required for a green transition. In many cases, IMF-mandated austerity measures limit fiscal space for climate adaptation and mitigation, while energy sector reforms focus on subsidy removals without clear commitments to reinvest savings into renewable energy or social protection.

Finally, we show attempts at *organizational experimentation* within the IMF. This was the case for the IMF's economic surveillance activities (known as Article IV consultations), which carry less policy weight than the IMF's lending, as implementation of the policy prescriptions is at the discretion of governments. Because there is less directly at stake in these activities, IMF staff have been advocating for a range of climate-related policies in their policy proposals.

This includes measures to mitigate transition risks, advice on carbon taxes, and advocacy for climate-related financial stress tests. However, even these measures are promoted against a backdrop of fiscal consolidation recommendations, which are unlikely to help countries implement their long-term climate objectives.

These findings directly relate to a burgeoning field of scholarship on how international organizations seek to 'mainstream climate.' Recent research shows that an ever-greater number of organizations are expanding their areas of operation into work that tackles climate concerns (Dellmuth et al. 2018; Dellmuth and Gustafsson 2021; Dellmuth, Gustafsson, and Kural 2020; Hall 2015; Persson and Dzebo 2019). For example, organizations as far-ranging as the World Bank, the United Nations High Commissioner for Refugees, and the World Health Organization have all been striving to integrate climate into their activities. In turn, such change has the potential to spur engagement and innovation on environment-related topics in broader organizational settings (Dörfler and Heinzel 2023). For the IMF in particular, scholarship has documented the emergence and intra-organizational spread – albeit uneven – of climate considerations in the organization's research and discourse (Clift 2024; Skovgaard 2021).

Our findings extend this line of scholarship by moving beyond analyses of organizational discourse or research to examine actual practices. Such a consideration draws on a long line of organization studies research that has shown plentiful instances of 'decoupling' between narratives and practices of organizations (Meyer and Rowan 1977). Our results document that the process of incorporating green transition objectives into organizational activities has been multifaceted and – at times – contradictory. We do *not* find mere ceremonial inclusion of green policies, but a case of rebuilding the ship at sea: the IMF has had to adapt and 'green' its policy activities at the same time as lending to countries in need and while the nature of appropriate climate policies remains contested. This has led to haphazard engagement with the issue, and a bias in favor of only one set of solutions to climate problems: those that support market mechanisms (such as carbon pricing) and market expansion (such as de-risking for green financing).

Ultimately, we seek to answer questions on *what* the IMF has been doing vis-à-vis climate and *how* it is doing it. What we do not seek to do is offer comprehensive answers to the questions of *who* has been behind these processes and *why*. These are analytically distinct issues, which we can turn to broader scholarship on the IMF and beyond to unpack. We do this in the concluding section and provide pointers for subsequent research on these topics.

2 Organizational Innovation: Introducing the Resilience and Sustainability Facility

2.1 New Money for Climate Change Adaptation and Mitigation

Established in 2022, the Resilience and Sustainability Facility (RSF) – a type of lending agreement – is the IMF's most significant foray into incorporating climate considerations into lending operations, as well as the only recent instance of a multilateral institution committing fresh financing for green objectives. Its introduction was the outcome of both organizational evolution and heated negotiations internally at the IMF and among its member-states. On the organizational side, the RSF represents the culmination of gradual recognition of the macroeconomic relevance of climate change, and the ensuing creation of various working groups and monitoring frameworks (Gallagher, Rustomjee, and Arevalo 2024). On the political side of its establishment, the pivotal moment was the fresh issuance of $650bn Special Drawing Rights (SDRs) – a kind of international reserve asset that countries can hold – in August 2021. These were allocated to each member-state in proportion with the size of their economies, meaning that high-income countries received the bulk of this new issuance. As they were less likely to need it to bolster their reserves, many high-income countries pledged to redirect $100bn into a new trust that would provide financial support to developing countries facing major risks due to climate change or future health emergencies. To this end, in October 2021, the G20 encouraged countries "with strong external positions to [...] voluntarily channel part of the [newly] allocated SDRs to help vulnerable countries" (G20 2021). Of this ambition, $46.7bn was pledged up to November 2024.

What makes these funds distinct from other IMF lending facilities is that they are intended to support policy reforms that reduce macroeconomic risks due to climate change over the long term. This contrasts the IMF's more traditional programs that have reform ambitions that only extend over the medium term (commonly, over the three to four years of their duration). To signal commitment to reform, RSF loan recipients need to simultaneously have in place a conditionality-carrying IMF program,[2] as well as sustainable debt burdens. In addition, the RSF loan also carries conditionality that borrowers must implement before any disbursements are made. In the first two years of its operation

---

[2] In some instances, these programs are not among the more common lending facilities, like Stand-By Agreements, External Fund Facility, or Extended Credit Facility. This was the case for Paraguay's RSF, which was joined with a program without credit (known as a Policy Coordination Instrument), and Jamaica and Morocco's RSFs are attached to low- or no-conditionality carrying programs (Precautionary and Liquidity Line and Flexible Credit Line, respectively).

(2022–2024), twenty-one programs were approved, a sign of the urgency of securing low-cost finance to support green transition objectives.

Given that the RSF entails fresh funds and new policy instruments for the IMF to translate its views on climate-sensitive economic policy into action, this lending facility offers a clear case of *organizational innovation* within global economic governance. This is multilateralism at work, relying on both the technocratic expertise of IMF staff in designing reforms and the political commitment of many of the organization's member-states that contributed new money to the facility. It also offers a unique type of support: financing for countries implementing comprehensive macroeconomic transformation programs that are intended to bolster their resilience in relation to climate risks, rather than project financing for specific climate change adaptation or mitigation projects – the latter of which falls under the remit of the World Bank and other development banks.

Further, the IMF's RSF stands out in the broader field of global climate finance, as it provides financing to climate-vulnerable countries on advantageous terms. For example, unlike the new Just Energy Transition Partnerships (JETPs) that offer funds by consortia of high-income countries on a mix of concessional and market terms and have been criticized for overly promoting creditor priorities (Bradlow and Kentikelenis 2024), the RSF is a truly multilateral alternative for financing green transition objectives. Consequently, closely tracking its practices and performance allows us to monitor whether the IMF is indeed helping to augment the policy space of low- and middle-income countries to finance green transition policies, or whether it is simply repackaging market-oriented policies in a new guise.

2.2 The Promise and Practice of Resilience and Sustainability Facility Loans

The raison d'être of the RSF is the provision of fresh funding to climate-vulnerable developing countries that lack the fiscal space to invest in climate change adaptation and mitigation. The facility's financing base is the SDRs that many high-income countries decided to re-channel into what became the Resilience and Sustainability Trust, which holds the resources that are used by its homonymous lending facility. Indeed, the IMF was quick to act on its establishment, partly reflecting the strong demand for such lending. The initial estimates of the organization suggested that eventually there would be about 33 RSF programs active in any given year (IMF 2022k).

The stated goals of the trust and the RSF are the support of reforms that reduce long-term macro-critical risks and "[a]ugment policy space and financial buffers" to mitigate these risks (IMF 2022k, 8). To meet these goals, financing was designed to have a 20-year maturity and 10.5-year grace period, to allow more fiscal space to countries implementing such reforms. Unlike the IMF's other concessional lending window, the Poverty Reduction and Growth Trust, eligibility for RSF loans is much wider: three-quarters of all IMF member-states (i.e., 143 countries) qualify for access to this financing source. Importantly, RSF programs cannot be stand-alone: a key eligibility requirement is that they complement one of the IMF's more traditional lending programs, most of which carry extensive conditionality. This means that borrowing countries still have to implement multiple reforms to satisfy the requirements of the concurrent programs.

The twenty-one loans that were approved in the first two years of the RSF's operations totaled SDR 7.3bn of committed resources.[3] This is higher than the pre-pandemic average of lending under the Poverty Reduction and Growth Trust, the IMF's main lending window for low-income countries, which averaged SDR 1.2bn per annum over 2000–2019 (IMF 2022a), although RSF financing has only partially overlapping criteria with this Trust. Codifying the early experience with RSF lending, the IMF issued an operational guidance note for staff in late 2023, where the organization clarifies that RSF conditionality – known as Reform Measures – should focus on "reforms that lead to permanent institutional changes, such as by involving legislative changes" and should have a catalytic effect that would facilitate private investment and even "help attract private investors" (IMF 2023g, 20, 24). Such explicit links to private finance – including through consultations between IMF staff and investors to "identify reform measures that can reduce barriers to climate finance" and "reduce the risk of such investments" (IMF 2023g, 29, 30)– form a novelty for IMF lending programs, which are generally focused on macroeconomic policies rather than involved in microeconomic decisions of firms.[4] The rationale, according to the IMF, is that only through such mobilization of private finance can adequate amounts be raised for climate change adaptation and mitigation, as most climate objectives "are not attainable without the mobilization of significant private finance" (IMF 2023g, 24). However, recent scholarship has criticized this approach for limiting the possibility of green developmental states emerging, instead relying on states to assume the risks of private investment (Gabor 2021).

---

[3] The recipient countries as of December 2024 include Bangladesh, Barbados, Benin, Cabo Verde, Cameroon, Costa Rica, Cote d'Ivoire, Jamaica, Kenya, Kosovo, Madagascar, Mauritania, Moldova, Morocco, Niger, Papua New Guinea, Paraguay, Rwanda, Senegal, Seychelles, and Tanzania.

[4] However, the IMF's guidance note also specifies that the IMF "cannot seek to mobilize climate financing by development banks or investors on behalf of the member, nor act as a financial advisor, and it cannot be involved in the management or oversight nor vouch for the bona fides or success of any climate finance vehicle or fund, or any climate project. The Fund focus is to provide policy advice to support an enabling environment for productive investments" (IMF 2023g, 30).

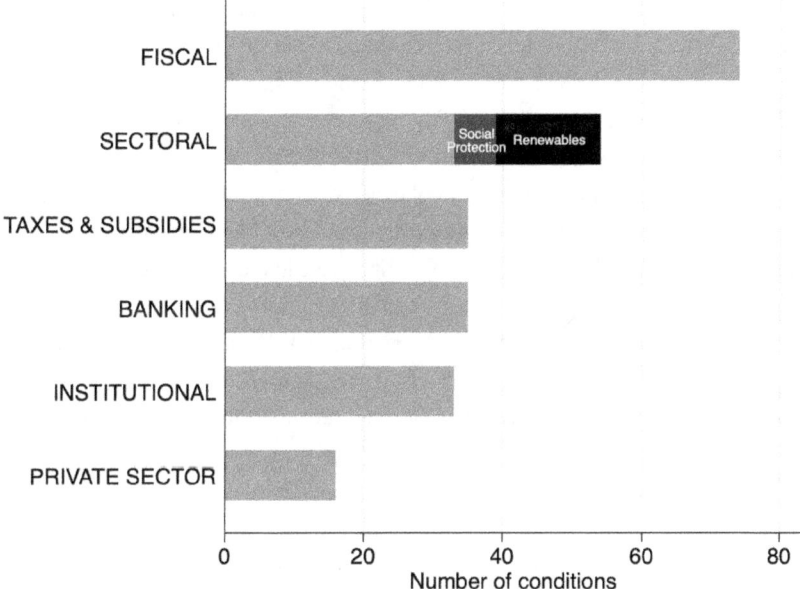

Figure 2 RSF condition policy areas

Source: Authors.

To investigate the actual content of the IMF's RSF programs, we collected all Reform Measures specified in the twenty-one agreements – yielding a total of 247 conditions – and then classified them into mutually exclusive policy areas. As Figure 2 shows, nearly a third of Reform Measures targeted fiscal policy: these reforms were almost always related either to Public Financial Management or Public Investment Management measures vis-à-vis climate policy. For example, Benin and Rwanda committed to implement 'climate budget tagging' to ensure climate is embedded into domestic budgetary decision-making, while Mauritania committed to incorporate climate considerations in all aspects of its public investments.

Turning to sectoral policies, this is a varied set of Reform Measures aiming at altering policy in economic sectors of relevance to the IMF borrower. For example, Benin, Morocco, and Barbados had to implement reforms to water use, Benin and Seychelles on revamping building codes, and Kosovo and Paraguay on electricity markets. Importantly, about a quarter of sectoral Reform Measures directly targeted renewable energy. Paraguay implemented reforms to enhance use of electric vehicles and incentivize recycling, Kosovo made steps to increase use of wind and solar power in its energy matrix, and Jamaica introduced new fiscal incentives for investment in renewables. Finally, six of the fifty-four sectoral measures directly related to social protection issues, and five of those sought to institute or reform social registries, a measure related to the introduction of compensatory measures for energy subsidy removals (discussed in Section 2.3).

Banking issues, taxation and subsidy measures, and institutional reforms each accounted for approximately 13–15% of RSF conditions. In these instances, countries are required to implement policies like climate incorporation into bank stress testing, removal of energy subsidies, and the introduction of various legal frameworks to embed consideration of climate risks into different policies. Finally, RSF agreements also included measures pertaining to private sector investments into green transition policies (6% of total conditions). These were primarily of two types. First, they related to country efforts at mobilizing climate finance, as was the case in Cabo Verde and the Seychelles. Second, several Reform Measures sought to revamp domestic frameworks on public-private partnerships to include climate requirements, as was the case in Bangladesh, Jamaica, and Cabo Verde.

2.3 Case Study Evidence of Recent RSF Loans

The evidence on how RSF agreements incorporate climate into the IMF's activities in its first two years of operation is revealing, but also incomplete if examined in isolation: these reforms form only part of the broader apparatus of conditionality included in IMF programs. Borrowing countries still need to implement a range of additional reforms attached to their parallel conditionality-carrying IMF programs. The organization provides this financial support through a variety of lending instruments, ranging from the short-term Stand-By Agreements that are usually heavy on fiscal consolidation measures but generally only last twelve to eighteen months, to more holistic programs under the Extended Credit Facility (ECF) and the Extended Fund Facility (EFF) that have more leeway for mandating 'structural reforms,' including in relation to climate issues. The latter programs have medium-term horizons, and it is thus no surprise that sixteen of the twenty-one countries with RSF loans in our sample have parallel lending agreements under the ECF and/or the EFF. It is through a joint analysis of the parallel RSF and non-RSF conditions that we can generate a more complete understanding of how the IMF's overall policy advice for climate vulnerable countries has evolved.

To examine these issues, we focus on the experiences of two countries: Kenya, as a highly populous lower-middle-income country with high needs for climate change adaptation projects, and Senegal – also a lower-middle-income country – which offers an opportunity to explore the IMF's approach toward climate change mitigation, given the country's recent oil discovery. Both countries are evaluated poorly in relation to their climate vulnerability: Kenya ranks 145th and Senegal 144th out of the 183 countries assessed by the Notre Dame Global Adaptation Initiative (University of Notre Dame 2023).

Before elaborating on the case study evidence, we outline the main comparative findings of our analysis vis-à-vis climate-related aspects of these programs. First, a common feature of the two programs was the IMF's advocacy of extensive austerity measures. These were promoted despite the organization's acknowledgment that fiscal consolidation could have adverse effects on countries' abilities to invest in climate-proofing their economy. Second, both the Kenyan and Senegalese lending agreements promoted a broader set of policies that sought to alter countries' economic policy profiles and the relative role of the state versus the market in shaping development policies. Third, the elimination of fossil fuel subsidies was also a key element of the IMF's fiscal advice, along with their replacement by targeted social assistance measures.

2.3.1 Kenya's 2023 Loan

Kenya was among the first countries to receive an RSF loan. In July 2023, the IMF approved a twenty-month loan for $551mil, designed to complement the country's existing $2.3bn Extended Credit Facility and Extended Fund Facility (ECF-EFF) loan that was in place since 2021. The main focus of the original program was to reduce debt vulnerabilities through multi-year fiscal consolidation measures, strengthening governance arrangements, and enhancing the monetary and financial sector framework (IMF 2021f). This was a domestically controversial program, as it was briefly interrupted in early 2022 due to opposition surrounding reforms mandating new tax measures amounting to 0.4 percentage points of GDP, including income tax changes, excise rate increases, and value-added tax exemption removals – some of which parliament elected to amend or remove (IMF 2022i). Against this backdrop, the 2023 RSF component of the IMF's lending to Kenya aimed to incorporate climate risks into fiscal planning and the investment framework, reduce emissions through carbon pricing, enhance existing frameworks to mobilize climate finance, and strengthen disaster risk reduction and management.

The IMF recognized that despite Kenya's modest contribution to global greenhouse gas emissions, its economy is highly vulnerable to climate change shocks. In recognition of these challenges, the RSF contained nine conditions: introducing carbon taxes, providing green fiscal incentives in land and water management, regulating electricity markets to promote energy efficiency, developing a green finance taxonomy, issuing guidelines for financial institutions on reporting climate-related risks, using climate change scenarios in fiscal risk analysis, adopting climate-informed investment selection criteria, implementing climate budget tagging, and adoption of a framework to enable dissemination of a digital early warning system platform. While the ECF-EFF program contained no conditions that directly addressed climate-related challenges, several have indirect implications for a green and just transition.

Despite recognizing that increased public spending on climate adaptation and mitigation can improve Kenya's debt profile, the IMF had nonetheless been calling for an extensive multi-year fiscal consolidation since the ECF-EFF program began in April 2021. The program initially called for a decline in the primary balance from a deficit of 4.0% of GDP to a 0.2% surplus by mid-2024 (IMF 2021f). In this context, the government revised domestic fuel and energy prices to ensure fiscal consolidation consistent with the program objectives. Fuel subsidies were eliminated in March 2023 and cost-based upward adjustments to electricity prices were effective April 2023 (IMF 2023f). By the most recent review, the IMF was calling for a 1.7% surplus by the 2024–2025 fiscal year, totaling a major 5.7 percentage point increase to the primary fiscal balance over the course of the program (IMF 2024f).

Notable in its absence was any explicit consideration of the trade-offs involved of such fiscal consolidation measures in achieving climate objectives. The ECF-EFF program also appeared to be at cross-purposes with the RSF program: while the IMF notes lack of funding for adaptation projects when describing climate finances, it does not recognize that the very fiscal austerity they endorse may undermine the ability of the government to fund such projects. To address these gaps, the IMF's programs anticipated private climate financing would be scaled up. Kenya requires an estimated annual climate financing of about 6% of GDP to meet its updated Paris-aligned climate goals (Government of Kenya 2020). According to IMF analysis, only about one-third of Kenya's financing gap is being filled annually and the IMF has recognized that current private sector commitments are falling far short of projected requirements and – to the extent that such funds are provided – they are mostly channeled towards renewable energy projects and none towards adaptation efforts (IMF 2023f).

To attempt to spur further investment, the IMF's conditionality called for the introduction of new regulations on the electricity market, bulk supply, and 'net metering' (i.e., the ability of consumers to generate their own electricity). These measures were intended as signals to private investors to invest in certain sectors, such as electric vehicles and renewable energy, and pave the way for private sector generation companies to transmit electricity to their customers via bilateral power purchase agreements in deals that bypass the partly state-owned enterprise Kenya Power and Lighting Company (KPLC), which operates most of the electricity transmission and distribution system. Under the ECF-EFF, the

IMF also set a condition requiring the transfer of KPLC's power transmission lines to Kenya Electricity Transmission Company, that KPLC enter a commercial contract with Renewable Energy Corporation for future Rural Electrification Schemes maintenance costs, and for the establishment of a new governance structure that gives private shareholders better representation by end-December 2024.

2.3.2 Senegal's 2023 Loan

Following on from an IMF lending program for $650mil that was approved in 2021, Senegal agreed on renewed financial support for three years in 2023. This was to be disbursed through a joint ECF-EFF program ($1.51bn) and a parallel RSF loan ($324mil) – a financing arrangement that mirrors Kenya's experience described earlier. The stated aims of the ECF-EFF are to reduce debt vulnerability through fiscal consolidation (primarily via revenue mobilization and phasing out energy subsidies), to strengthen public sector governance, and to foster a more inclusive and private-sector-led growth trajectory (IMF 2023i). The aim of the RSF is to support efforts to tackle climate change mitigation objectives, accelerate climate change adaptation, and integrate climate change considerations into the budget process (IMF 2023i).

The RSF program contained ten conditions grouped into three pillars. The first pillar seeks to support climate change mitigation goals by adopting an implementation plan for greener public transport and phasing out untargeted subsidies in the electricity sector. The second pillar aims to accelerate adaptation to climate change by adopting a country development strategy consistent with its Nationally Determined Contributions, submitting urban planning laws that mitigate coastal erosion and urban flooding, disseminating climate risk data to relevant committees, and strengthening the institution in charge of implementing water management. The third pillar looks to integrate climate change considerations into the budget process by integrating climate considerations into public investment management, assessing disaster-related fiscal risks for the 2025 budget law, developing guidelines for climate budget tagging and issuing directives in the budget call circular to reflect climate investments, and ensuring all public investments and public-private partnership projects include an appraisal of their effect on climate adaptation and mitigation.

The selection of these reforms was coordinated with the World Bank, who joined the program negotiation mission and are providing ongoing technical assistance for reforms required under the RSF (IMF 2023h). Extensive coordination is also evident with regard to the policy measures mandated in the World Bank's Development Policy Financing program commencing May 2022 that are consolidated in the ECF-EFF and RSF programs, such as reforms geared towards establishing the regulatory framework for private investment in the gas subsector and increasing transparency in government procurement processes in the electricity sector (World Bank 2022).

Like in the case of Kenya, several conditions related to the reduction of the fiscal deficit have the potential to impact the country's climate change efforts. The IMF calls for a fiscal consolidation that would see a decline in the primary budget deficit from 4.4% of GDP at end-2022 to 0.7% by 2025. Fiscal consolidation is to be driven primarily by a phasing out of untargeted energy subsidies – via reductions to gasoline and diesel subsidies under the ECF-EFF and via increases to electricity tariffs under the RSF – from 4% of GDP in 2022 to 1% of GDP in 2024, and to eliminate them by 2025. To facilitate this process, the IMF set structural conditions under the ECF-EFF requiring the government to establish an independent committee with the mandate to determine and publish final consumption fuel prices based on a revised pricing formula; and to revise the current pricing formula for petroleum products to ensure that prices at the pump reflect developments international markets by December 2023. The IMF also included a condition under the RSF stipulating that the government adopt an electricity tariff adjustment plan that would see subsidies gradually eliminated. These reforms hold implications both in terms of the shift away from dependence on fossil fuels and the extent to which this shift is consistent with a just transition. The IMF argues that reductions to energy subsidies effectively raise the price of fossil fuels to the end-user, thereby encouraging less and more efficient use of energy and incentivizing a shift to cheaper renewable sources, like wind and solar (Lagarde and Gaspar 2019).

In the context of austerity measures and energy price hikes, the IMF included some provisions for the protection of vulnerable groups. For example, the program mandated the re-certification of data in the Single National Registry and a scaling up of its coverage from 500,000 to 1,000,000 households to improve targeting of the most vulnerable individuals in the safety net system, while the government was also expected to settle outstanding cash transfers that had not been paid since 2021, which "would therefore help protect the most vulnerable in the context of energy subsidies phasing out" (IMF 2023i, 20). More specifically linked to a just green transition, IMF conditionality mandated that the new electricity tariff structure incorporates a social tariff specifically designed for vulnerable groups. It is as yet unclear to what extent this may alleviate the burden on vulnerable households, reflecting a more general trend throughout the program documentation of a lack of detailed diagnostics surrounding the impact of energy subsidies on poorer income deciles and substantive analysis indicating how social protection measures will compensate. In addition, it is questionable whether the IMF's push for the rapid elimination of energy subsidies is politically feasible given Senegal is in a presidential election year and has already experienced social unrest – and is only likely to be exacerbated by, in the IMF's words in describing the Senegalese context, "a cost-of-living crisis, widespread youth unemployment, and threat to security from developments in neighboring countries [that] are adding to this loaded political period" (IMF 2023i, 5).

2.4 Summary

Since its launch in 2022, the RSF has rapidly gained traction, with twenty-one approved programs in its first two years. Unlike traditional IMF lending, the RSF explicitly integrates climate considerations into macroeconomic policymaking, providing long-term financing on favorable terms. However, access to RSF funds is contingent on countries having an active IMF program with structural reforms, meaning that climate financing is intertwined with broader economic policy conditions. Our analysis of RSF loan agreements reveals that conditionality primarily focuses on fiscal policy, sectoral reforms, and mobilizing private finance for climate objectives. While this approach aims to strengthen resilience, it also raises concerns about the extent to which IMF-supported policies expand policy space for climate action versus reinforcing pre-existing market-oriented policy frameworks.

The case studies of Kenya and Senegal illustrate the potential and limitations of RSF lending. While both programs seek to enhance climate adaptation and mitigation strategies, they are embedded within broader IMF arrangements that emphasize fiscal consolidation. The IMF's assumption that private sector mobilization can fill climate financing gaps remains largely untested, and there is limited evidence that such strategies will enable a just and effective green transition. Furthermore, the tension between austerity measures and climate investment needs remains unresolved, raising questions about the IMF's role in shaping the climate finance agenda. Thus, while the RSF represents an important organizational innovation in global economic governance, its long-term impact will depend on whether it genuinely expands fiscal space for climate action or primarily serves to reinforce existing IMF policy priorities.

3 Organizational Inertia: The IMF's Standard Lending Agreements

3.1 The Resurgence of IMF Lending after the Pandemic

Although the Resilience and Sustainability Facility (RSF) represents a major organizational innovation for the IMF, most of its lending still takes place outside the confines of this new climate-oriented lending facility. During the acute phase of the COVID-19 pandemic, no- or low-conditionality facilities were the organization's most prevalent lending instruments, but these were gradually replaced by traditional lending arrangements – like the Stand-By Arrangement or the Extended Credit Facility – that mandate the introduction of policy reforms over a period of one to four years. As shown in Table 1, between 2020 and 2024 the IMF approved sixty-four traditional IMF condition-carrying programs, providing access to SDR 109bn in credit. The majority of approved loans, both in terms of quantity (forty-six programs) and total resources provided (SDR 94bn), are *not* linked to the RSF. Importantly, many of these non-RSF borrowers are also countries with extensive climate vulnerabilities: twenty-nine programs are with countries ranked in the bottom third of the Notre Dame Global Adaptation Initiative (2023) index that measures vulnerability to and readiness for climate shocks. This reveals the central role of the IMF in setting macroeconomic policies of countries affected by climate change and the green transition the most and, by extension, helping determine the resources that countries have at their disposal to pursue adaptation and mitigation strategies.

While the IMF's own 2018 review of program design identified gaps in conditionality related to climate change issues (IMF 2019a), the organization has since emphasized that policy reforms attached to its programs have fundamentally changed to take climate risks more seriously. This follows on from its unambiguous recognition of climate change as central to its core mandate of supporting and safeguarding macroeconomic stability, with the 2021 Climate Strategy clarifying that financing may be provided "when climate-related measures are deemed critical to solve a member's balance of payments problems" (IMF 2021d). In the case of traditional lending programs, the stated aim is to help build 'climate-resilient' economies, not least because this will help countries pre-empt major balance of payments disruptions, the raison d'être for IMF interventions.

In this section, we analyze data from thirty-three non-RSF-linked loan agreements active between March and June 2024 (the point of data collection) alongside case studies of Argentina and Pakistan. Our findings reveal *organizational inertia* within the IMF, as its standard lending programs, despite actively shaping the policy environments of borrowing countries, have remained resistant to change. Rather than adapting to incorporate climate considerations meaningfully, these programs continue to follow the IMF's established modus operandi. This suggests that efforts to mainstream climate into the organization's operations have been inconsistent and inadequate. As a result, most

Table 1 List of IMF programs approved between January 2020 and December 2024

	2020		2021		2022		2023		2024	
	Number of countries	SDRs	Number of countries	SDRs	Number of countries	SDRs	Number of countries	SDRs	Number of countries	SDRs
Traditional programs **without** RSF loan	7, of which 3 are highly climate-vulnerable	13.7bn	9, of which 7 are highly climate-vulnerable	5.6bn	9, of which 3 are highly climate-vulnerable	42.5bn	12, of which 9 are highly climate-vulnerable	19.9bn	9, of which 7 are highly climate-vulnerable	12.5bn
Traditional programs **with** RSF loan	None	n/a	5, of which 3 are highly climate-vulnerable*	5.3bn	4, of which 2 are highly climate-vulnerable*	1.4bn	8, of which 6 are highly climate-vulnerable	7.3bn	1, which is highly climate-vulnerable	0.3bn

* Programs are listed as having an RSF component even before the RSF started lending in 2023 because several RSF loans were added on to pre-existing IMF programs approved before 2023. That is, the 'traditional programs with RSF loan' category classifies as starting year the onset of the traditional IMF program, not the onset of the RSF add-on.

Notes: Loans include the Extended Credit Facility, Extended Fund Facility, Stand-By Arrangement, and Stand-By Credit Facility. We exclude agreements under the Flexible Credit Line, Precautionary and Liquidity Line, and Short-Term Liquidity Line because they are low- or no-conditionality programs. SDRs reported on RSF-linked programs exclude the RSF portion of the program. Climate vulnerability based on ND-GAIN index ranking, where we classify rank 125 to 187 as high. See Online Appendix for full list of programs.

countries subject to IMF-mandated reforms still face policy prescriptions that may run counter to the objectives of a green transition.

3.2 Macroeconomic Adjustment and the Role of the Energy Sector

This section assesses how traditional IMF lending programs consider climate issues and the green transition. To do this, we collected lending agreements of the thirty-one countries that had an ongoing IMF program between March and June 2024 that were *not* joined with an RSF loan; these were the most recently available data at the point of data collection. From these reports, we extracted a battery of information on the treatment of climate risks and considerations in different policy recommendations, the inclusion of climate-related conditions, and the nexus between climate and fiscal concerns.

3.2.1 Fiscal Consolidation and the Green Transition

The global green transition will demand large-scale investments from various sources. While the exact mix of public, private, and multilateral financing for this goal varies by country, there is wide-ranging acceptance that the role of states is central, both as funders and as regulators (Mazzucato et al. 2024; Rodrik 2014). But this role comes under pressure over the course of IMF interventions due to the organization's demands for extensive fiscal consolidation measures. To a degree, changes to spending in countries undergoing deep economic crises – the most common reason for resorting to the IMF in the first place – are inevitable. What matters most is the scale, scope, and pace of such measures: how much will be cut, in which areas, and how fast? It is in these regards that the IMF has a central role in shaping country decisions through the conditions attached to its loans, which specify a wide range of quantitative fiscal targets, as well as many structural reforms to help attain these targets.

Evidence on the IMF's recent programs reveals that austerity measures form a key part of its policy advice to countries. Of the thirty-one countries with recent IMF programs, twenty-eight had available data on fiscal consolidation, and budget cuts were foreseen for twenty-four of these countries, averaging contractions to the tune of 4.2% of GDP.[5] Disaggregating these IMF borrowers by their income classification reveals that low-income countries (nine countries) will bear the brunt of austerity, with average IMF mandated contractions of 4.9% of GDP, while middle-income borrowers (fifteen countries) are

---

[5] The period and duration of fiscal consolidation vary across programs in our sample. Program start dates range from March 2021 to June 2024, and the duration for consolidation is between 1 and 4.

required to contract by 3.7%. A main reason for these high levels of fiscal austerity is to repay debt: according to the IMF and World Bank's latest assessment of debt sustainability, thirteen out of the nineteen contracting countries for which we have data are classified as being 'in distress' or 'at high risk of distress.' As a country in debt distress, Zambia provides a case in point. Following a major drought, the COVID-19 pandemic, a debt default, and spillovers from the Russian war in Ukraine, the country commenced an IMF program in 2022 that mandated extensive fiscal consolidation: 5.0% of GDP over three years (IMF 2024l). These cuts occurred in a context of urgent financing needs for climate adaptation, as the country is experiencing an increase in frequency and severity of droughts and floods (Ngoma et al. 2021).

Of course, austerity measures have been part of the IMF's toolkit for decades (Kentikelenis and Stubbs 2023), so their promotion in recent lending programs should not come as a surprise. But this time is different because such measures are in part defended through a climate rationale: a key plank of such fiscal consolidation is the decrease in fossil fuel energy subsidies to consumers and/or producers. Indeed, price increases or subsidy cuts in electricity, gas, or fuel were requested in twenty-two of thirty-one countries with recent IMF loans. For example, Suriname's IMF program necessitated a fiscal consolidation of 5.5% of GDP over 2021–2025, of which 3.9% of the savings are from the reduction of energy subsidies (IMF 2024j). As a result, fuel subsidies were eliminated in March 2023; liquefied petroleum gas subsidies were reduced to 55% of the cost, resulting in a price increase of 425% in September 2023 (with further increases slated); and average electricity tariffs increased in the second, third, and fourth quarters of 2023 by 28%, 42%, and 21% respectively, with the aim of eliminating subsidies (cost-recovery) by end 2024.

Deep, far-ranging, and fast-paced austerity measures can undermine green transition objectives. First, and most conspicuously, reductions in public spending impact available financing for green investments: as governments seek cuts, future-oriented projects – like infrastructural upgrading or renewable energy – are often the first to be retrenched (Stubbs and Kentikelenis 2023). In other words, short-term fiscal goals trump concerns over long-term sustainability. Second, austerity diminishes the ability of states to recruit, train, and retain qualified personnel (ActionAid 2021), which in turn reduces their regulatory and enforcement capacity (Reinsberg et al. 2019). This opens the door for entrenched interests to block or stall major green initiatives and for weak enforcement of climate-related regulations. Third, the social backlash from austerity – stemming from citizens' exposure to reductions in the availability of social services or the introduction of new taxes – erodes the potential support base for green policies, insofar as these are perceived to require public financing or additional spending by households (Beiser-McGrath 2022; Beiser-McGrath and Bernauer 2019).

Regarding the latter, the IMF also points to how such reforms are implemented with simultaneous concern for their distributional implications. For this reason, the organization promotes compensatory measures – commonly in the form of so-called 'social spending floors' that stipulate a particular level below which social spending should not sink during the IMF's intervention. Of the twenty-two countries that the IMF calls for decreases in energy subsidies, twenty-one contained nonbinding quarterly indicative targets on priority social expenditures, and thirteen contained explicit reference to compensatory measures for households in direct relation to higher energy prices. For example, the Democratic Republic of Congo's loan called for a 7% increase in fuel prices in April 2024, which brought total increases since the program began to over 75%. The IMF loan agreement stipulated that the authorities should "ensure a timely phasing out of direct fuel subsidies, to reduce tax expenditure (implicit subsidies), and develop targeted social safety nets for vulnerable populations" (IMF 2024b, 13). Similarly, the IMF's approach to a just transition in Chad's loan was focused on ensuring the phase-out of subsidies is coupled "with appropriate transfers to households to mitigate the impact on vulnerable ones" (IMF 2023a, 20).

These examples typify the IMF's approach to compensation for the removal of subsidies: targeted transfers to the most vulnerable groups, underpinned by social spending floors. But the way the IMF implements this strategy often presents several issues. First, the IMF neglects the sequencing of reforms. Compensatory mechanisms take time to build, and key populations that would benefit from compensation can struggle to access these resources (e.g., due to bureaucratic hurdles, complex eligibility criteria, or inadequate information). Yet, the IMF often expects cuts to subsidies from the outset of the program, *before* mechanisms have been put in place to protect vulnerable households. For instance, the IMF advised Gabon in July 2022 that "[fuel] subsides should end by end-December 2022," despite Gabonese authorities' indicating that "their option of protecting all households [via fuel subsidies] is based on the absence of well-targeted social safety nets" (IMF 2022e, 13). While the IMF also urged authorities to "finalize the database of the most vulnerable by end-December 2022, and subsequently develop well-targeted measures" (IMF 2022e, 13), even the most optimistic timeline would leave vulnerable households exposed for a six-month period while subsidies are being phased out but no targeted measures are in place. Only in Honduras were sequencing issues explicitly mentioned, where the IMF stated that "authorities should consider reversing the reductions to gasoline and diesel taxes introduced in 2022 – at an annual cost of 0.4% of GDP – *once the social safety nets have been further strengthened*" (IMF 2023d, our emphasis, 15).

Second, a widely advertised feature of IMF programs that purportedly helps protect social groups impacted by the economic fallout of the organization's lending programs is the aforementioned social spending floors, whose functioning and application were codified in the Strategy on Social Spending (IMF 2019b). The promise of these measures is that they allow – or compel – countries to protect or increase public investment in health, education, and social protection. But recent evidence points to such floors being inconsistent, inadequate, and not implemented (Kentikelenis and Stubbs 2024).

Third, there is a lack of transparency in terms of how the resources freed up from energy subsidy elimination are repurposed. Savings could be directed toward increases in social protection and investment for renewable energy, but this is difficult to verify due to the insufficient provision of data in IMF program reports detailing where such savings will be used. For example, in the Republic of Congo's loan, the IMF called for "reprioritiz[ing] resources freed up by the elimination of fuel price subsidies (2.7 percent of non-oil GDP for 2023) to social spending targeted at vulnerable groups (0.4% of non-oil GDP, including measures targeting public transportation) and capital spending (1.5% of non-oil GDP), providing a positive impulse to inclusive growth while still achieving fiscal consolidation" (IMF 2024h, 11–12). Notably, the increase in capital spending includes agriculture, roads, electricity, health, education, transportation, as well as promoting tourism, industry, the digital economy, and special economic zones, but no indication is provided for how spending is distributed among these sectors (and with little guarantee that it was not redirected to supporting the country's hydrocarbon industry).

3.2.2 Green Structural Adjustment in the Energy Sector

The discussion has thus far presented summary evidence of the IMF's advice on fiscal policy and social protection for its recent borrowers. These are features of nearly all traditional IMF programs. However, beyond IMF targets on public expenditures, lending programs also include a range of structural reforms – that is, measures that target deep changes to a country's policy environment and that include fine-grained interventions on policy areas considered by the IMF as important.[6] Such IMF reforms vary considerably in each country, and, per Table 2, we distinguish its borrowers between established fossil fuel producers (either major global producers, or producers for whom fossil fuel rents form a

---

[6] Our use of the term "green structural adjustment" differs from that of Bigger and Webber (2021). They refer to the World Bank's efforts to restructure local governments in low- and middle-income

Table 2 Recent IMF borrowers by fossil fuel production status

Category	Definition	Countries
Established fossil fuel producers	Either producing over 0.05% of global oil, gas, or coal in 2023, or having fossil fuel rents that exceed 3% of GDP	12 countries: Argentina, Chad, Rep Congo, Ecuador, Egypt, Gabon, Ghana, Pakistan, Papua New Guinea, Serbia, Suriname, Ukraine
Emergent fossil fuel producers	New fossil fuel projects are mentioned in the IMF review and the country is not already an established fossil fuel producer	3 countries: Mozambique, Somalia, Uganda
Nonproducers of fossil fuels	Countries not in the two groups above	16 countries: Armenia, Burkina Faso, Burundi, Central African Rep, Comoros, Dem Rep Congo, Gambia, Georgia, Guinea-Bissau, Honduras, Jordan, Malawi, Nepal, Sri Lanka, Togo, Zambia

Notes: Established fossil fuel producers defined as countries that are either producing over 0.05% of global oil, gas, or coal in 2023 or that have fossil fuel rents that exceed 3% of GDP annually from 2015 to 2019. Emergent fossil fuel producers coded as countries for which fossil fuel projects are mentioned in IMF program documentation (and where the country is not already an established fossil fuel producer). The remaining countries are classified as not being major producers of fossil fuels.

significant part of economic activity), emergent fossil fuel producers (countries with ongoing fossil fuel exploitation projects), and the remaining countries which are either nonproducers or where such production forms only a minor part of economic activity. If the IMF lives up to its own rhetoric on supporting the green transition, then its advice for the former two categories should be

countries and develop capacity to capture financial flows for investment in climate adaptation projects.

geared toward facilitating the decarbonization of these economies and against further extraction. Further, for fossil fuel producers and nonproducers alike, investments in renewable energy should be expected to form a major part of IMF policy advice.

Among the fifteen recent IMF borrowers that are either established or emergent fossil fuel producers, the IMF discusses macroeconomic effects of continued production through its analysis of ongoing reforms in eight of them. The rationale for covering extractive policies in the emergent or established fossil fuel producers frequently relates to fiscal or debt considerations. For example, in the case of Chad, the IMF explains that additional oil revenue will be central to "reduce domestic arrears and external debt" (IMF 2023a, 1). Per the IMF's advice, "higher oil revenue should be used to build buffers and accelerate the payment of debt. ... Debt is expected to decline faster than envisaged at the time of the [original request for the lending agreement], as higher oil prices would result in accelerated repayments to Glencore, while higher net oil revenue will allow for a quicker reduction of the stock of T-bills and payment of domestic arrears" (IMF 2023a, 15). This illustrates how debt problems can end up entrenching high-emission growth models, a phenomenon known as the debt-fossil fuel trap (Woolfenden 2023b). Similarly, in the case of Uganda, the IMF outlines how investments in the oil sector would improve the fiscal deficit (IMF 2024k). And in Gabon, the IMF explains that the government should "enhance facility maintenance and plan adequate investment for refurbishment and encourage exploration and discovery of new oil" to safeguard fiscal revenues and exports (IMF 2022e, 44). Even in Zambia, a country that is not a significant fossil fuel producer, the IMF promoted the construction of the Tazama pipeline, which is expected to reduce transportation costs, but will also lock in such fossil fuel investments over the medium run (IMF 2024l).

One possibility for the organization would have been to couple such recommendations with the promotion of investments in renewable energy, to gradually wean domestic energy production off fossil fuels. This was the case in only one of these eight countries: Ecuador. The country was advised to prioritize "unlocking private-led investment projects in renewable energy sources such as solar and wind, for which Ecuador has good potential" (IMF 2024d, 18).

The IMF fared little better in its attention to such issues for nonproducers of fossil fuels. Of the sixteen such recent borrowers, the organization treated renewable energy issues in three countries. This advice was geared toward supporting the role of the private sector in production and distribution processes, as the IMF's analysis in two of these three countries referred to measures that would expand markets. Such measures have potential for spurring investment in renewable energy, but they can also foreclose more strategic and centralized approaches that accord states a more active role in steering such transformations. For instance, Jordan adopted an implementation roadmap for the electricity sector that would "facilita[te] a further shift to renewable energy sources and an increase in competition" (IMF 2024e, 53). In Georgia, the development of a renewable energy support scheme will see the government award 1,500 MW of power generation capacity to the private sector via competitive auctions (IMF 2022f). While market competition has the potential to lower prices and displace the dominance of fossil fuel usage, it can also lead to uncoordinated efforts and a short-term focus on profits rather than on investments that will ensure viability in the long run (Gabor and Braun 2025).

The role of private sector promotion in recent IMF programs raises broader questions about the possible biases that the organization's policy advice has. The pro-market approaches that the organization promotes are well established, even though – in recent years – there has been cautious acknowledgment on the potential of state-led programs for green transformation. For example, recent IMF research has noted the potential for industrial policies to increase domestic renewable energy manufacturing capacities (Prasad et al. 2022). Indeed, nearly 30% of all industrial policies that have been developed in 2023 have been justified by governments on climate grounds, thus prompting the IMF to point to the return of industrial policy (Ilyina, Pazarbasioglu, and Ruta 2024).

However, the evidence from the organization's recent lending programs reveals that engagement with the potential for green industrial policies was nonexistent. The closest the IMF came to advocating such policies was in the Democratic Republic of Congo, where the organization acknowledged the potential for becoming a major player in the global green transition, given its abundant mineral wealth, and the importance for "expanding along the mining value chain" (IMF 2024b, 55). Even so, the practical steps promoted were primarily regulatory interventions: improvements in mining governance and transparency, and enforcement of social and environmental regulations. While such measures are important, they do not constitute major efforts to develop comprehensive green industrial policies that would ensure broad-based economic development.

Instead, the IMF set conditions that called for state-owned enterprise restructuring and deregulation that would expand the remit of private actors in the energy sector. In this regard Serbia stands out. The IMF targeted Elektroprivreda Serbije, the country's coal-dominated electricity utility, mandating the adoption of a comprehensive restructuring plan that specified "reform priorities in the areas of organizational and financial restructuring, human resources, procurement, project development, reporting, risk management and the environment" (IMF 2024i, 13). Similarly, Georgia was expected to drive reforms to its major state-owned enterprises, including the Georgian State Electrosystem and the Gas Transportation Company (IMF 2022g).

3.3 Case Study Evidence of Recent IMF Loans

In this section, we focus on recent experiences of two countries with traditional IMF loans: Argentina, which demonstrates the IMF's approach to climate change mitigation in the context of newly discovered oil and gas reserves; and Pakistan, which illustrates how the IMF navigates climate change adaptation. Both countries are established fossil fuel producers and among the top 30 for global greenhouse gas emissions, therefore representing critically important cases.

Pre-empting our findings, the cases show that IMF lending programs affect borrower countries' capacity to facilitate a green transition and fulfil climate adaptation objectives above all via fiscal policy. In Argentina, energy subsidy reductions were consistent with a transition away from dependence on fossil fuels, but the encouragement of private sector energy investment to limit reliance on energy imports and generate export earnings was not. The needs of the poorest were safeguarded to some degree, as they were protected from energy subsidy reductions, and the envisaged domestic fossil fuel production may offer reprieve from high energy prices. In Pakistan, removal of tax breaks on renewable energy technologies represented a fundamental threat to the country's green transition. And while energy price increases and subsidy reductions could aid decarbonization, they were not politically or socially palatable, despite attempts to buffer the poorest households with an expansion of social support schemes. Ultimately, both programs lacked a serious consideration of climate-related objectives, despite clear relevance to the IMF's package of reforms and in terms of the physical risks to livelihoods and transition risks to the economy presented by climate change.

3.3.1 Argentina's 2022 Loan

Argentina is the twenty-third largest emitter in the world, contributing 0.8% of global greenhouse emissions in 2021 (World Resources Institute 2024a), and contains the second largest reserve of shale gas and the fourth largest reserve of shale oil worldwide (IEA 2024a). Its energy mix is dominated by fossil fuels, of which 55% is from natural gas and 33% from oil (IEA 2024a). Despite pledges to engage in renewable energy projects, the government is ramping up gas and oil extraction by heavily subsidizing natural gas exploration in the untapped deposits of Vaca Muerta in Northern Patagonia, with the goal of exporting 50% of its crude oil and 38% of gas (Fundación Ambiente y Recursos Naturales

2019). The government also subsidizes fossil fuels to the end-user. In 2021, energy tariffs reflected 37% of the full cost of electricity and 44% for gas, and were among the lowest in Latin America, while subsidies amounted to 2.3% of GDP (Parry, Black, and Vernon 2021). Moreover, the country is a large agriculture emitter, linked to its dependence on large-scale livestock and agricultural industries, with meat and grain exports accounting for 55% of export revenues (World Bank 2021a).

Argentina entered a thirty-month $44bn IMF program in March 2022 to replenish foreign exchange reserves, control inflation, reduce the budget deficit, and improve long-term growth (IMF 2022b). The country was in the midst of a full-blown economic and social crisis since 2018, exacerbated by the COVID-19 pandemic and the war in Ukraine. An unsustainable public debt burden and current account deficit had drained the country of foreign exchange reserves, and a collapse of investor confidence meant there was little access to new loans. Inflation had reached over 50% year-on-year, fueled by the Central Bank printing money to support the budget deficit as well as global commodity price rises, placing a disproportionate burden on the poorest households. Most significantly, the new IMF loan meant Argentina could avoid defaulting on debt repayments of $19bn in 2022 owed to the IMF itself because of a previous lending program in 2018.

Argentina's IMF program affects three key arenas in relation to the country's climate change efforts: energy subsidy reforms, energy investment, and identification of climate risks. First, energy subsidy reforms were mandated in IMF conditions related to fiscal consolidation that would see a decline in the primary budget deficit from 3.0% of GDP in 2021 to 0.9% by 2024. To reach these targets, the program called for reducing the energy subsidy bill by 0.6% in 2022, with further reductions scheduled for subsequent years. These reforms can support climate objectives by raising the price of fossil fuels to the end-user, thereby encouraging more efficient usage of energy and incentivizing a shift to renewable sources like solar. However, the IMF discussed energy subsidies only in relation to immediate fiscal risks, thereby overlooking opportunities to foster a green transition in the long term – for example, by reorienting savings from fossil fuel subsidies toward green transformation. While the removal of energy subsidies can place a disproportionate burden on poorer households, these are being phased out in a progressive way: eliminated for the top 10% of urban residential users and for large commercial users, pegged so that prices reflect 40% of average wage growth for poor households, and pegged so that prices reflect 80% of average wage growth for other residential users. This indicates the IMF's constructive role at designing effective approaches in phasing out energy subsidies.

Second, Argentina's IMF program encouraged the development of a strategic tradeable sector in energy to replenish foreign exchange reserves and ensure long-term debt sustainability. To this end, the IMF endorsed the government promoting private investment in exploration, production, and transportation of energy from the shale oil and gas reserves of Vaca Muerta. The IMF described how "efforts are needed to improve the efficiency of the energy sector and tap Argentina's vast energy potential to reduce reliance on more expensive energy imports and transition to a cleaner energy matrix" (IMF 2022b). In this context, the 'cleaner energy matrix' refers to natural gas, which is described as a transition fuel in Argentina's Nationally Determined Contributions. Ultimately, it would allow the country to become an exporter of natural gas and reduce its reliance on expensive energy imports. But such incentives for fossil fuel expansion contradict the IMF's advertised climate orientation that centers on aiding countries phasing out of fossil fuel usage. It represents a missed opportunity to consider large-scale government investment in renewable energy production and supply-side subsidies to encourage private investment in wind and solar. Furthermore, where consumer energy subsidies are being cut at the behest of fiscal expediency, it could be viewed as contentious to advocate incentives for wealthy private investors – including TotalEnergies, Wintershall Dea, and Pan American Energy (Fundación Ambiente y Recursos Naturales 2021) – through producer subsidies, tax breaks, and other guarantees. Even so, Vaca Muerta represents the potential to reduce domestic energy costs, including for poorer households, and can also offer a crucial source of foreign exchange that would enable the country to avoid painful IMF austerity programs in the future.

Third, the IMF considered physical risks linked to climate change, such as climate-induced commodity export shocks impacting production and exports, but provided only negligible coverage of such risks. The organization also failed to identify global spillover transition risks linked to the government's economic dependence on Vaca Muerta energy reserves and on highly pollutive agroindustry. As an increasing number of countries commit to decarbonization, trade partners may impose carbon border taxes, impacting the earnings from such exports. There was also no recognition of national-level transition risks linked to asset stranding in the fossil fuel sector if the country chooses to move toward an energy matrix dominated by renewables.

3.3.2 Pakistan's 2020 Loan

Pakistan contributed 0.9% of global greenhouse gas emissions in 2021, ranking as the twenty-first largest emitter (World Resources Institute 2024d). Roughly two-thirds of its energy needs are met by fossil fuels, including oil (25%), natural gas (24%), and coal (14%), and about 40% is imported (IEA 2024d), which depletes foreign exchange reserves and exposes the country to inflationary pressures due to international fossil fuel price shocks. Pakistan faces significant physical risks from climate change, foremost of which is the greater frequency and intensity of flooding, especially along the Indus River, where most of the population resides. Floods have cost the economy an estimated $1.7bn annually since 2010, and damages are projected to increase to $5.8bn per year by 2030 (World Bank and Asian Development Bank 2021b). Climate change is also increasing the incidence of severe droughts, disproportionately affecting the livelihoods of the 39% of the workforce engaged in agriculture, most of whom are subsistence farmers (World Bank and Asian Development Bank 2021b). Heat waves, too, are expected to increase in frequency and intensity, resulting in reduced crop yield and desertification, critical water shortages, floods from rapidly melting glaciers, and economically debilitating power cuts (Aftab 2022; Ellis-Petersen and Baloch 2022).

The IMF resumed a thirty-nine-month $6bn lending program with Pakistan in March 2020 – it was initially approved in July 2019 but interrupted due to the pandemic. The resurrected program sought to reduce the fiscal deficit, control inflation, replenish foreign reserves, and improve the financial viability of the energy sector (IMF 2021g, 2022j). It responded to urgent balance of payments issues that arose due to a surge in the value of imports linked to higher commodity prices, which depleted Pakistan's foreign exchange reserves.

Pakistan's IMF program impacts three important facets of the country's climate change efforts: tax reforms, energy sector reforms, and identification of climate risks. First, tax reforms were predicated upon a series of IMF conditions to achieve a fiscal consolidation of 2% of GDP for the 2022 fiscal year. As a result of the removal of various goods from tax zero-rating, discounted rates, and other exemptions, a 12% increase in sales tax was implemented for imported electric vehicles and a 20% tax was introduced on solar panels, wind turbines, and other renewable energy technologies (Moulvi 2022). By increasing the price of renewable energy, the IMF-mandated tax reforms disincentivize investor uptake, including from existing fossil fuel-based producers, self-generating agriculturalists, and industrial consumers. Major beneficiaries of the growth in solar and wind energy in Pakistan have been poorer subsistence farming communities that remain without grid access, so by placing higher costs onto these communities – which are also the most impacted by climate change – the tax reforms undermine achievement of a socially just low-carbon transition.

Second, Pakistan's program contained conditions to reduce the fiscal deficit through energy subsidy and pricing reforms as part of a more comprehensive energy sector restructuring. The energy sector suffers from long-standing deficiencies related to circular debt – a complex form of public debt accruing due to unpaid government subsidies to distribution companies, who in turn cannot pay independent power producers, thereby affecting the entire power and gas chain. Two reforms co-occurred that raise costs to the end-user but reduce government expenditures: energy *price* reform, which seeks to bring electricity and gas prices in line with cost recovery along prescribed formulas and procedures; and energy *subsidy* reform, which seeks to target subsidies to a smaller group of consumers on a more progressive tariff structure. These energy sector reforms could support climate objectives by raising the price of fossil fuels to the end-user, thereby providing more incentive to invest in energy-efficient production capacity or shift to off-grid renewable sources like solar. However, more ambitious reforms to the energy sector were overlooked because the IMF's recommendations were guided by short-term fiscal expediency, rather than considering climate concerns that will impact macroeconomic fundamentals in the long run. For instance, cheaper renewable energy could allow for a financially viable long-term solution to addressing the sector's recurrent deficiencies. Furthermore, fuel and power price hikes disproportionately burden poorer households because energy constitutes a greater portion of their spending – and energy prices in Pakistan were already three times higher than other countries in the MENAP (Middle East, North Africa, Afghanistan, and Pakistan) region (Parry, Black, and Vernon 2021). Recognizing this issue, the IMF called for vulnerable households to be compensated via expansion and better targeting of social support schemes like the Benazir Income Support Program. However, recent social tensions over food and energy prices suggest that this package of measures is neither politically palatable nor sufficiently compensatory as poorer households struggle to free up resources to adapt to new economic realities and the consequences of climate change (Ethirajan 2023; Mangi 2024).

Third, Pakistan's IMF program contained negligible coverage of physical or transition risks. The IMF identified climate change as a risk to the program, but the analysis was too vague to be of substantive use. Higher frequency and severity of natural disasters is simply flagged as a potential cause of severe economic damage. Outside of the risk assessment, the IMF described the recent history of extensive damage that climate-related disasters have dealt to the Pakistan economy, and the country's position as one of the world's largest emitters on an absolute basis. These descriptions were self-contained and did not consider trade-offs and contradictions involved between the program and Pakistan's extensive climate adaptation and mitigation needs. For instance, the IMF advised the country to accelerate efforts to meet greenhouse gas reduction commitments by reforming energy prices, subsidies, and taxes, but failed to recognize how fiscal constraints prescribed by the program impede efforts to invest in new infrastructure and incentives for climate adaptation. In addition, the IMF did not consider risks linked to the banking sector from changes in carbon-intensive asset values or include any climate-related stress tests in its debt sustainability analysis. As a result, the IMF failed to quantify the macroeconomic benefits of climate adaptation and mitigation measures.

3.4 Summary

This section revealed organizational inertia in the IMF's standard lending programs vis-à-vis their approach to climate change. Despite the Fund's stated commitment to mainstreaming climate considerations, our analysis of thirty-three non-RSF-linked loan agreements and case studies of Argentina and Pakistan suggest that its traditional lending programs remain largely resistant to change. These programs continue to prioritize fiscal consolidation, deregulation, and market-oriented reforms, often at odds with the investments and policy shifts required for a green transition. In many cases, IMF-mandated austerity measures limit fiscal space for climate adaptation and mitigation, while energy sector reforms focus on subsidy removals without clear commitments to reinvest savings into renewable energy or social protections. The IMF's push for private sector-led energy transitions also risks sidelining state-led green industrial policies, which are crucial for long-term sustainability and are now widely practiced in high-income countries.

The case studies further illustrate the contradictions in the IMF's approach. In Argentina, while energy subsidy reductions align with decarbonization goals, the IMF's support for fossil fuel expansion in Vaca Muerta undermines these efforts. In Pakistan, tax reforms that increased the cost of renewable energy technologies directly discouraged the country's transition to a low-carbon economy. Across both cases, climate risks – whether from fossil fuel dependence or physical climate shocks – were acknowledged but not meaningfully integrated into macroeconomic policy design. Overall, the evidence suggests that the IMF's standard lending operations continue to reinforce rather than reform the existing economic structures that contribute to climate vulnerability, raising concerns about whether the institution's broader climate strategy is truly transformative or merely superficial.

4 Organizational Experimentation: Economic Surveillance and Climate Advice

4.1 The Role of Economic Surveillance in IMF Operations

A core but underappreciated area of the IMF's operations is the surveillance of its members' economic and financial policies. This is primarily achieved through the research and analysis underpinning the organization's 'Article IV reports,' which are published annually for large economies and a bit more infrequently for smaller ones. In addition, a supplementary economic surveillance tool, the Financial Sector Assessment Program (FSAP), is jointly organized with the World Bank and aims to identify financial sector vulnerabilities as well as opportunities to contribute to development objectives. For forty-seven IMF member countries with systemically important financial sectors, it is mandatory to participate in a regular FSAP: thirty-two countries are evaluated once every five years (including several large developing countries, like China, Brazil, India, and Mexico), and once per decade for fifteen emerging markets (IMF 2021c). In this section we focus solely on the Article IV consultations, which also incorporate findings from the FSAPs.

Even though they lack the element of compulsion that lending agreements have, the IMF's surveillance activities are still highly significant. First, domestic policymakers are informed of the IMF's assessment of their country's economic outlook and the range of policy advice that IMF staff offer. This advice does not come with any enforcement capacity, but it is nonetheless consequential for countries, especially those at lower levels of economic development, where such advice shapes policy debates. Indeed, this group of countries form the majority of IMF borrowers and would-be borrowers, and the analyses and advice contained in the Article IV report commonly informs the reforms mandated in IMF lending programs. In contrast, high-income countries often disregard the advice offered in these reports, as they are seen as incompatible with the practical and political constraints that these countries face (Edwards and Senger 2015; Momani 2006).

Second, as Article IV reports are made public and IMF Executive Board members are invited to discuss their findings, this process also serves to transmit important information between countries (Schäfer 2006). For example, critical Article IV reports can yield peer pressure for countries to pursue reforms that will contribute to regional or global financial stability. Or donor countries might rely on these reports to learn more about the policy environments and challenges of potential aid recipients. Such information-sharing via the IMF forms a key source of knowledge on the economic conditions of other countries around the world.

Finally, the IMF's economic surveillance impacts financial markets and investor decisions. Article IV reports give signals to international capital markets on whether a country's economic policies are credible, in turn affecting the availability and cost of credit. Indeed, financial markets have been shown to react favorably to the publication of positive press releases on surveillance findings (Fratzscher and Reynaud 2011). Similarly, private actors use Article

IV reports to make decisions on where to direct investments or to inform their own forecasts of economic performance of countries around the world (Frenkel, Rülke, and Zimmermann 2013). This is especially the case for low- and middle-income countries, where private investors are typically less informed about economic developments (Breen and Doak 2023), and where extensive economic policy research capacity is often lacking (Fellesson 2017; Sanyal and Varghese 2006).

4.2 Climate Change Policies and IMF Surveillance

Despite their influence, until recently Article IV reports – and the related FSAPs – neglected coverage of the economic impact of climate change, including a range of physical risks from natural disasters, transition risks due to the shift to a low-carbon economy, and spillover risks on account of the economic fallout of a green transition in a country's major trading partners (Gallagher et al. 2021; Ramos et al. 2022a; Volz and Ahmed 2020b). Even by the IMF's own admission, climate-related analyses were haphazard (IMF 2021a), while civil society also called out the organization's delayed action on this front (Kentikelenis and Stubbs 2021a, 2021b; Sward et al. 2021).

To address these shortcomings and criticisms, the IMF announced an overhaul of surveillance practices in the 2021 Comprehensive Surveillance Review (IMF 2021a, 2021b). The Review commits the IMF to adopt a systematic approach to integrating climate-related risks into its bilateral surveillance activities, in line with the high-level commitments by Managing Director Kristalina Georgieva (Georgieva 2021; IMF 2020b). Accordingly, climate change mitigation measures should be covered at least every three years for the twenty largest greenhouse gas emitters; risks in the transition to a greener economy should be assessed for economies dependent on fossil fuel production, including coverage of revenue and expenditure policies as well as the broader set of regulatory or institutional reforms that can aid this objective (e.g., the need for fossil fuel exporters to diversity their export base); and climate change adaptation and resilience analysis should be undertaken in countries that are especially vulnerable to climate shocks (for a list of sixty-four countries, see IMF 2019c).[7] Building on these commitments, in 2022 the IMF published a Staff Guidance Note that offered broad guidelines for operational purposes (IMF 2022h), as summarized in Table 3.

---

[7] Coverage of the latter is not entirely new, with Christine Lagarde making progress during her leadership on considering how adaptation issues link to debt sustainability, especially for small-island economies and natural disaster-prone countries (Gallagher, Rustomjee, and Arevalo 2024).

Table 3 Coverage of climate change in Article IV consultations

	Mitigation	Domestic transition management	Spillover transition management	Adaptation
What does it mean?	Contributing to the global mitigation effort	Achieving domestic emission targets	Adjusting to the global low-carbon transition	Building resilience against effects from climate change
Why cover it in Article IV consultations?	Spillover effects to other countries	Domestic policy challenge	Domestic policy challenge and spillover effects arising from external shocks or policy actions in other countries	Domestic policy challenge and spillover effects arising from external shocks or policy actions in other countries
Which countries should it be covered in?	Strongly encouraged for 20 largest emitters, encouraged for others	Only where it significantly affects present or prospective balance of payments or domestic stability.	Only where it significantly affects present or prospective balance of payments or domestic stability	Only where it significantly affects present or prospective balance of payments or domestic stability, especially the most vulnerable countries

Source: Adapted from IMF (2022h, 54).

These developments represent a sea change compared to the broadly unchanged operations of IMF bilateral surveillance over the years. Reflecting this, the IMF's Executive Board emphasized that "implementing the new modalities ... revolve around the principle of experimentation" (IMF 2021b, 3). Not only does this indicate the organization's broader acceptance that climate change issues are inherently critical for macro-economic performance (or 'macro-critical' in IMF parlance) but it also reflects an exercise in *organizational experimentation* as the IMF seeks to reposition itself as a key actor facilitating the green transition for all 191 member countries (and not just those facing pressing balance of payments issues).

4.3 Case Study Evidence of Recent IMF Surveillance

The promise of the IMF's organizational experimentation into climate-sensitive economic surveillance is clear, but what does recent evidence reveal? To what extent is policy advice consistent with enabling countries to transition away from dependence on fossil fuels? Does such advice address transition risks and financing needs arising from shifts to renewable energy? In the remainder of this section, we analyze recent IMF Article IV reports for three countries: Indonesia (IMF 2023e), South Africa (IMF 2022l), and Colombia (IMF 2023b). Indonesia and South Africa stand out as two upper-middle income G20 nations that are among the largest twenty greenhouse gas emitters and rely heavily on fossil fuels for energy and to generate foreign exchange reserves and government revenues from exports (Arinaldo and Adiatma 2019; Rumble and Sidiropoulos 2022). Colombia, another upper-middle-income country, is selected given both the importance of fossil fuels and extractive industries to its economic development model and its high ambitions for an energy transition since the presidential election victory of leftist Gustavo Petro (Bocanegra 2022). For all three countries, global efforts to phase out fossil fuels will directly impact their energy mix and export markets and fundamentally alter the economic prospects and livelihoods of their inhabitants, thus presenting crucial cases for an assessment of current practices in IMF surveillance. We do not select an adaptation-centric case given the IMF's focus on large emitters in economic surveillance.

Anticipating our findings, the cases reveal evidence of organizational experimentation into climate-sensitive economic advice. In Indonesia, the IMF highlighted deficiencies in its carbon tax scheme while deploying economic models to illustrate the macroeconomic costs of meeting domestic climate change commitments; in South Africa, domestic transition risks linked to employment opportunities and to the carbon-intensive activities of its state-owned enterprises were scrutinized; and in Colombia, spillover risks were examined using economic simulations and underpinned advice on the need for export diversification. The IMF also frequently conducted climate-related stress tests on the financial sector and recognized the need to protect poorer households to facilitate energy subsidy reform.

Nonetheless, there were several recommendations that could hamper the achievement of green transition objectives. Most notably, the IMF's analyses in all three countries favored fiscal consolidation without adequately considering how it might impact medium- and long-term climate strategies, including the associated risks from not investing in adaptation and mitigation. Evidence from past bouts of rapid fiscal consolidation have shown that it can have adverse follow-on implications for economic growth and business activity (Blyth 2013; Dreher 2006; Ortiz and Cummins 2021; Ostry, Loungani, and Furceri 2016; Przeworski and Vreeland 2000), which in turn limit the capacity of households and the private sector to adapt to and mitigate climate change. Further, the Article IV reports devoted only ad hoc coverage of physical, transition, and spillover risks, rather than integrating them into a systematic analytical framework that traces their effects on the economy. Climate change issues were also compartmentalized rather than embedded in the general analysis of economic prospects and challenges. Despite these limitations, climate-related analytical work in the IMF's Article IV reports has clearly become more sophisticated compared to past highly cursory treatment of these issues.

4.3.1 Indonesia's 2023 Article IV Consultation

Indonesia boasts the largest economy in Southeast Asia, underpinned by an export sector led by coal (18.5% of 2022 exports) and palm oil (9.0%) (OEC 2024). It is the sixth largest emitter in the world – contributing 3.1% of global greenhouse gas emissions in 2021 (World Resources Institute 2024c) – and is the world's fourth largest coal producer, largest gas supplier in Southeast Asia, and meets 77% of domestic energy needs with fossil fuels (IEA 2024c). Its government supports the fossil fuel industry via a range of subsidies, at 15.4% of GDP in 2022 (Black et al. 2023), while the fossil fuel sector provides a major share of the government's revenues, accounting for 13.6% of total revenues over 2014–2016 (Braithwaite and Gerasimchuk 2019). Indonesia is also one of the planet's largest emitters of land-use related emissions, an outcome of deforestation and peatland fires undertaken to facilitate the expansion of oil palm plantations (World Bank and Asian Development Bank 2021a).

Given Indonesia's significance in the fight against climate change, the IMF's Article IV analyses offered in-depth coverage of green financing issues, which entailed mobilizing private investment to finance Indonesia's adaptation and mitigation commitments. Further, the organization welcomed the Indonesian government's introduction of a carbon tax of 30,000 Rupiah (about $2) per ton of carbon dioxide equivalent that will apply to coal-fired power plants and came into effect in 2022. The IMF demonstrated consideration of climate mitigation in identifying key limitations of the scheme. For instance, it recognized that since the government provides energy subsidies and sets the price for fossil fuels and electricity, the carbon tax will not provide an incentive for end-users to transition to renewable energy and/or achieve greater energy efficiency. Further, the IMF supported the 2021 tax reform law to raise additional revenue, which included broadening of excise taxes to plastics.

The IMF also endorsed cuts to energy subsidies to help balance the budget. The Indonesian government sets fuel and electricity prices for consumers below market rates and compensates producers for the difference. As global oil prices surged in 2022, spending on consumer subsidies and compensation for producers tripled (an increase of about 2 percentage points of GDP). The IMF recommended changes to the pricing formula that would align electricity and fossil fuel prices with the market price, viewing such reforms as both fiscally expedient and "essential to change incentives in the energy sector and help achieve climate objectives" (IMF 2023e, 12). The rationale is that it forces energy end-users to internalize the full cost of fossil fuels. While increases in energy prices disproportionately impact poorer households (Nasruddin 2022), the IMF explicitly recognized the need to expand social assistance benefits and coverage to facilitate energy subsidy reform.

In addition, IMF staff referred to earlier analyses on the macro-criticality of climate change and transition risks the country faces (IMF 2021e). As multilateral and private banks and investment managers commit to coal divestment, this major export of Indonesia may face lower demand, which would impact the viability of coal companies, the domestic energy mix, and – by extension – the broader economy. The organization also provided analyses using the global IMF-ENV model (Chateau, Jaumotte, and Schwerhoff 2022) – which captures detailed sectoral, trade, and employment consequences of mitigation policies to address climate change – to illustrate the macroeconomic costs of meeting domestic climate change commitments based on several scenarios, using the evidence to bolster calls for the country to undertake carbon pricing and energy subsidy reform.

However, the Article IV report also contained recommendations that could hinder Indonesia's green transition. On the fiscal policy side, the organization advocated for a budget deficit ceiling of 3% of GDP by 2023, despite the financing needed to achieve Indonesia's Nationally Determined Contribution targets alone amounting to 2.8% of GDP annually (Government of Indonesia

2022). Given the IMF's assessment of the Indonesian economy as having "ample policy space, strong financial buffers, and favorable initial conditions to respond to adverse shocks" (IMF 2023e, 20), there was policy space for greater ambition vis-à-vis climate. An unduly cautious fiscal approach could represent a threat to Indonesia transitioning away from fossil fuel dependence and achieving its climate commitments, as investment on climate adaptation and mitigation measures need to be scaled up. Relatedly, the organization did not adequately reflect on the role of carbon-intensive sectors like coal, oil and gas, and palm oil in driving higher-than-expected revenue performance. These sources of government revenue cannot be relied on in the long-term as Indonesia and its trade partners transition toward a low-carbon economy and may thus spell fiscal trouble in the medium- or long-term.

The IMF also called on the government to phase out nickel ore export restrictions and not extend them to other commodities. The government plans to diversify exports away from raw commodities by encouraging manufacturing, marketing, packing, and/or retailing to add value before reaching export. Thus far the government focused on nickel through tax holidays for investment in smelter capacity and the introduction of an export ban on raw nickel in 2020. The strategy proved economically successful: foreign direct investment from China and Hong Kong led to smelters increasing from three in 2014 to eleven in 2023, processed nickel exports surged from $4.5bn in 2019 to $19.6bn in 2022, and formal sector jobs were created in poorer regions such as Central Sulawesi and North Maluku (IMF 2023e). Following the experience with nickel, the government intends to extend the strategy to copper, bauxite, cobalt, and tin to promote manufacturing of batteries for electric vehicles. Moreover, the IMF failed to analyze the strategy's environmental costs, describing the main costs as "foregone fiscal revenues, the unintended consequences of export restrictions at home (such as potential resource misallocation and rent seeking), and those that spillover across borders (such as price effects in the global commodity markets), which could potentially be met by retaliation from trade partners" (IMF 2023e, 27). Yet, new coal power plants are being constructed to power nickel industrial parks and other metal smelters, which could threaten the achievement of climate change goals (Simon 2023).

The IMF's analysis contained several additional blind spots. In the debt sustainability analysis, IMF staff did not assess the long-term risks from climate change adaptation needs to public debt and gross financing needs (IMF 2023e), thus failing to quantify and convey the benefits of adaptation policy measures vis-à-vis the country's debt profile. In addition, the IMF did not consider global spillover risks linked to trade partners committing to decarbonization through the imposition of carbon border taxes or related measures, thereby impacting the potential earning from fossil fuel exports and environmentally unsound extractive sectors like palm oil. China, for instance, is the main importer of Indonesian coal and has already introduced a national carbon pricing mechanism (Nogrady 2021), which could decrease the country's demand for coal from Indonesia. Finally, more ambitious reforms to the energy sector were overlooked, despite their significant macroeconomic implications. For instance, the possibility of more affordable renewable energy sources could offer a financially sustainable, enduring remedy for the power sector's ongoing overcapacity issue from coal-fired power plants, which is costing the government $1.2bn annually in operating and maintenance expenses (Prasetiyo et al. 2023).

4.3.2 South Africa's 2021 Article IV Consultation

South Africa is the largest economy in Africa and the world's seventeenth largest emitter, with 1.1% of global emissions in 2021 (World Resources Institute 2024e). Over half of merchandise exports are from the carbon-intensive mining sector, including coal, platinum, gold, and iron ore (World Bank 2021b). Its economy is also one of the most coal-dependent in the world. The coal mining sector employs over 90,000 workers concentrated in regions with high unemployment levels and millions more are connected to the coal value chain, such as transport, electricity generation, and petro-chemical production (Rumble and Sidiropoulos 2022). Moreover, about 94% of South Africa's energy needs are met by fossil fuels, primarily from coal (71%) and oil (20%), and the country produces 84% of its electricity from coal (IEA 2024e). Furthermore, subsidies were handed out over the past decade to support the production and consumption of petroleum and coal. In 2019 alone, the government spent $4.3bn on fossil fuel subsidies (Climate Transparency 2021).

There were some climate-sensitive developments in South Africa's Article IV report coverage of climate change and the green transition. IMF staff highlighted that the room for active government support of climate adaptation and decarbonization transition had been constrained because many state-owned enterprises are highly exposed to carbon-intensive activities (e.g., coal-fired power plants, rail, and port infrastructure), which makes them vulnerable to a drop in demand from the decarbonization transition, with potential significant fiscal implications. On the other hand, IMF staff outlined how South Africa's jobless pandemic recovery – there were 1.9 million less people employed at the end of 2021 compared with the quarter before the pandemic (Rumble and Sidiropoulos 2022) – will mean that the migration of low-skilled workers out of the coal value chain will be even more challenging, and that deficiencies in the country's education system further complicate the necessary workforce transition. To address these issues, the IMF advised South Africa to improve the quality of education, apprenticeships, and vocational training schemes to support displaced workers, and to design policies that could bridge the spatial divide between workers' living areas and places where new jobs are created.

In addition, the IMF was constructively critical of South Africa's policies on climate grounds. In its assessment of the country's Economic Recovery and Reconstruction Plan (Government of South Africa 2020), the IMF identified important inconsistencies with the aim of a low-carbon economic rebound – for example, a published list of preferred bidders under the Risk Mitigation Independent Power Procurement Program showing that most of the two gigawatts of energy procured use carbon-intensive gas technology. The IMF also acknowledged that the banking and financial system faced significant physical risks related to natural disasters and transition risks related to coal-based energy generation, and their analysis included stress tests for how future climate-related policy developments might affect financial stability.

Despite these climate-sensitive steps, other areas of the IMF's policy advice could potentially hamper green transition objectives. Most notably, the IMF endorsed expenditure cuts to reduce the fiscal deficit from −3.9% of GDP in 2021 to −1.8% by 2023. Further, the IMF also did not evaluate the extent to which the fiscal consolidation may impede the ability of the government to scale up public investment to fulfil the climate adaptation and mitigation programs described in its Nationally Determined Contribution (Government of South Africa 2021).

In relation to South Africa's energy sector, the IMF argued that reforms to reduce rigidities in the economy are key to accelerate decarbonization of the power sector and transition away from coal. Eskom, the near-monopoly national power utility, faces persistent criticism for its outdated and poorly maintained coal power plants, which have led to economically debilitating rolling-blackouts – known locally as 'loadshedding' – since 2008 that continue to the present as it struggles to keep pace with growing demand (Cohen and Burkhardt 2022). The IMF criticized Eskom, which relies heavily on government transfers and coal extraction, because the company had actively resisted new entrants into the sector by delaying the expansion of independent power producer programs that would allow for the growth of renewables. Nonetheless, the IMF fell short of recommending additional incentives for investors to enter the renewable energy market or for new forms of large-scale government investment in renewables (i.e., distinct from Eskom). The IMF also identified a need to expedite the authorization process to accelerate significant investments by several mining companies to generate their own electricity, and to reduce regulatory hurdles and tackle a backlog of mining licensing applications to attract investment in the mining sector. The promotion of mining sector investment is clearly counter to a green transition and, as a carbon-intensive activity, there is imminent risk that such investments will become stranded assets. Additionally, if mining company electricity generation is based on fossil fuels, then such advice may further entrench fossil fuel dependence.

More generally, the IMF provided insufficient recognition of spillover risk to public finances and the balance of payments due to the ongoing low-carbon transition. South Africa is expected to lose $84bn by 2035 from falling prices and demand because of other countries' transitions to a low-carbon economy (Huxham, Anwar, and Nelson 2019). Investments by Eskom and other carbon-intensive sectors of the economy will also become exposed to asset stranding, with significant flow-on effects for the broader economy (Burton et al. 2016). This represents a major omission – especially so given that the IMF explicitly recognizes that South Africa's subpar economic performance over the last decade is the result of economic policies failing to adapt to the end of the commodity price boom of the 2000s. The impending drop in demand, and thus prices, for carbon-intensive commodities as South Africa's trade partners commit to decarbonization thus represents a level of urgency that warrants embedding in all projections and assessments of fiscal risk.

4.3.3 Colombia's 2023 Article IV Consultation

Colombia is the fifth largest economy in Latin America and the Caribbean, the world's thirty-first largest greenhouse gas emitter (World Resources Institute 2024b), and remains highly dependent on commodity exports. The country's crude oil and coal exports composed 55% of merchandise exports in 2022, while agricultural products like coffee, cut flowers, bananas, and palm oil represented 15% (IMF 2023c). About 73% of Colombia's energy needs are met by fossil fuels, primarily from oil (41%) and gas (23%) (IEA 2024b). But the Petro-led administration made ambitious plans for an energy transition. This included prioritizing the expansion of renewable energy production and diversification of exports to reduce the country's dependence on fossil fuels, as well as approving a progressive tax reform law that included an annual wealth tax, higher duties on oil and coal exports, a windfall tax for oil and gas companies, and an increase and broadening of the scope of the carbon tax (Deloitte 2022).

The Article IV report appropriately emphasized the need for export diversification. As key export markets shift toward renewable sources or consider carbon border taxes, the IMF reflected on the underlying reality that over-reliance on fossil fuel exports is unlikely to provide a reliable source of export earnings and fiscal revenue in the future. This advice was bolstered by economic simulations showing oil production declines of about 90% by 2033 would result in GDP dropping by 1.3%, a current account deficit ratio of 6% of GDP, and a fall in fiscal revenue by 2% of GDP. Moreover, the IMF recognized that dependence on fossil fuels is likely to increase in the coming years because of the tax reform, as more than half of the reform's total yield comes from taxes on the oil and coal sector.

Another area where the IMF offered value was through its FSAP (IMF 2022c), included as an annex to the Article IV report. A transition risk stress-testing exercise assessed the effects of a higher carbon tax on the banking sector at both a granular and aggregate level, based on increases of $10, $15, $20, and $70 per ton of carbon; and a physical risk stress test was performed at the municipal level to investigate banks' vulnerability to riverine floods. Based on these findings, the IMF suggested the government "adopt a risk-based approach in supervision for climate-related risks and continuously improve information disclosures (both by nonfinancial corporates and by financial institutions) and data availability" (IMF 2023b, 48).

A sticking point for the IMF was the government resources dedicated to fuel subsidies, estimated at about 2.6% of GDP in 2022 (World Bank 2023a). Gasoline and diesel price smoothing occurs in Colombia through a fuel price stabilization fund, effectively acting as an untargeted fuel subsidy. Domestic fuel prices were below international prices following price freezes during the first half of 2022, resulting in the fund accruing a deficit. The IMF endorsed the government's plan for cutbacks of fuel subsidies, which will allow domestic gasoline prices to be higher than international prices while focusing subsidies on diesel products that are typically consumed by poorer households (Government of Colombia 2023). This presents a step toward reducing the consumption of fossil fuels while also guarding against distributionally regressive consequences.

However, the IMF also advocated policies that could impede green transition objectives. The organization pushed for a fiscal consolidation that would see the government balance shift from a −1.1% of GDP deficit in 2022 to a 0.6% surplus in 2023, an adjustment larger than required to remain within limits prescribed by the country's fiscal rules (which the IMF helped design). At a time when expenditure on climate adaptation and mitigation needs to be scaled up, fiscal consolidation beyond what the country's fiscal rules necessitate could nullify new revenue sources for Colombia to invest in transitioning away from fossil fuels. For instance, the World Bank (2023a) estimates that the Colombian government will need to spend an additional 0.4% of GDP per year between 2023 and 2030 to fund its transition, based on the optimistic assumption that the private sector is able to provide another 1.1% of GDP. Yet the IMF fails to explicitly examine the implications of its fiscal targets for climate initiatives.

In relation to export diversification, the IMF called for "a market-oriented strategy to sustain the strong dynamism of non-traditional exports of recent years. Import substitution and protectionist measures should be avoided" (IMF 2023b, 16). Such advice is at odds with arguments by IMF staff who recognize the central importance of industrial policies to increase domestic renewable energy manufacturing capacities (Prasad et al. 2022). It also neglects Colombia's underwhelming experience with the role of private companies in the energy transition, which encounter strong opposition from local communities (Reuters 2023). Instead, industrial policies developed by the government in consultation with the relevant stakeholders can offer a more sustainable and equitable path forward. For example, the Ministry of Environment and Sustainable Development in Colombia is working on a *comunidades energéticas locales* project to enhance community participation in energy projects and to share in its revenues.

There were additional missed opportunities in the Article IV report. First, any energy market interventions – precisely because of their direct impact on the income distribution and their potential for political destabilization – should be accompanied by distributional impact assessments. The IMF has employed such a tool in the recent past – for example, it quantified how different economic policies would impact each income decile in the case of Ecuador (IMF 2022d). Yet such an analysis was altogether absent from the recommendations to Colombia. Second, the IMF recommended a tight monetary stance beyond 2023, including further hikes to interest rates. While this advice may be appropriate, the IMF failed to conduct analyses of how it will affect investment in renewable energy. Higher interest rates can slow the renewable energy transition and shield oil and gas producers from competition by low-carbon producers because newly applied renewable energy technologies have relatively large front-loaded costs compared to already-installed fossil fuel technologies (Ferguson and Storm 2023). Finally, the IMF failed to analyze needs related to managing the job transition, as opportunities linked to new export sectors may not geographically correspond with oil and mining areas.

4.4 Summary

This section explores the evolution of the IMF's economic surveillance practices, particularly focusing on how climate change considerations have been increasingly integrated into its Article IV consultations. Hitherto underemphasized, climate issues gained prominence following the IMF's 2021 Comprehensive

Surveillance Review, which mandates systematic incorporation of climate-related risks into its assessments. Through case studies of Indonesia, South Africa, and Colombia, we illustrate the IMF's mixed performance in aligning its economic policy advice with green transition goals. While the IMF has made strides by recognizing transition risks, advising on carbon taxes, and conducting climate-related financial stress tests, it has also been favoring extensive fiscal consolidation measures without adequately addressing their potential to undermine long-term climate objectives.

The IMF has been cautious in advocating for the kinds of expansive fiscal policies necessary for robust climate adaptation and mitigation investments. It also tends to compartmentalize climate risks rather than embedding them comprehensively into economic forecasts and policy recommendations. This compartmentalization, coupled with an inconsistent application of climate risk assessments across countries, suggests that while the IMF is experimenting with incorporating climate into its surveillance, it has yet to fully reconcile its traditional fiscal priorities with the urgent demands of the global green transition.

5 Conclusions: Real but Limited Organizational Change

5.1 Unevenly Green, Unflinchingly Market-Oriented

Policy change at the IMF vis-à-vis climate issues has been real but limited. The IMF has increasingly acknowledged the necessity of countries pursuing effective climate change and green transition policies, which not only make environmental good-sense but are also necessary conditions for macroeconomic stability in the medium- and long-run. However, the extent of change remains constrained by its long-standing preferences for deep and fast budget cuts and its consistent advocacy for market-based solutions to policy problems.

The introduction of the Resilience and Sustainability Facility (RSF) represents the clearest case of *organizational innovation* that integrates climate considerations into lending. However, the RSF remains a small fraction of total IMF financial commitments and is conditioned on adherence to broader structural reforms that often align with pre-existing macroeconomic stabilization policies.

The *organizational inertia* evident in the IMF's traditional lending programs – not linked to the RSF – further underscores the limits of change. The conditions mandated in these loan agreements continue to prioritize fiscal discipline and market-based solutions, often at odds with the fiscal expansion needed for green investments. As seen in the case studies on Argentina and Pakistan, IMF policies, despite their rhetorical support for climate adaptation and mitigation, reinforce fossil fuel dependence and impose fiscal constraints that hinder long-term sustainability initiatives.

Moving beyond the IMF's lending activities, the other main area of IMF country-facing operations – the surveillance of its members' economic policies – shows signs of *experimentation* with the incorporation of climate-related advice. These so-called Article IV reports have incorporated climate concerns more systematically, in line with high-level commitments by the organization. But this shift has been uneven: surveillance missions now frequently devote considerable attention to the fiscal and economic implications of climate change, including carbon pricing and transition risks, yet these discussions often coexist with recommendations that emphasize fiscal consolidation and market liberalization. As a result, while climate considerations have entered the IMF's analytical frameworks, they have not fundamentally reshaped the main thrust of its policy advice.

A central feature of the IMF's approach to the green transition is its emphasis on market-based solutions. Carbon pricing, subsidy reductions, and private sector mobilization are dominant elements in the Fund's climate policy recommendations, thus confirming arguments that the IMF is pursuing solutions to climate change that are overly focused on enhancing the role of private business (Gabor 2021). While the private sector has a key role in contributing to emission reductions, the effectiveness of such policies also depends on whether they are embedded in a broader policy ecosystem that supports the structural transformations necessary for a green transition. In other words, the role of the public sector is key in supporting and steering these policy changes, yet this has been consistently de-emphasized in the IMF's policy advice.

For instance, the IMF's focus on reducing energy subsidies is framed as both a fiscal necessity and a tool for discouraging fossil fuel consumption. In effect, we see a 'green' fiscal consolidation logic where austerity is defended through a climate rationale of decreasing public expenditure on fossil fuel energy subsidies. Yet, in many low- and middle-income countries, subsidy removals occur without the simultaneous development of adequate social safety nets or public investment in renewable energy infrastructure. This sequencing problem has led to social unrest in several countries, highlighting the sociopolitical risks of relying too heavily on market mechanisms without complementary state interventions. Similarly, the IMF's climate finance strategies assume that private investment will fill the gap left by constrained public budgets, even though evidence from RSF programs suggests that private sector mobilization remains limited, particularly in economies with weak financial institutions or extensive climate risks.

Our findings raise important questions about who is driving this organizational change and why. While a full treatment of these dynamics is beyond the scope of the present analysis, we can draw on related scholarship to sketch the broad contours of the actors involved. First, scholarship on international organizations highlights the pivotal role of senior management at the IMF in directing institutional priorities (Copelovitch and Rickard 2021). In particular, the tenure of Christine Lagarde as IMF Managing Director (2011–2019) is widely recognized as instrumental in propelling the Fund to ever-greater incorporation of climate considerations into its policies (Clift 2024; Gallagher, Rustomjee, and Arevalo 2024; Skovgaard 2021). This momentum persisted under Lagarde's successor, Kristalina Georgieva, who sought to mainstream climate concerns across the Fund's operational and analytical activities (Ramos et al. 2022b).

Second, the IMF's internal bureaucracy has also emerged as an independent driver of policy change, similar to the important role of the World Bank's bureaucracy (Heinzel, Weaver, and Jorgensen 2024; Lang, Wellner, and Kentikelenis 2024; Weaver et al. 2022). Recent evidence suggests that rotation practices among IMF staff expose them to countries with a diversity of climate vulnerabilities, and this exposure – in turn – shapes their perception of environmental risks and viable policy responses (Clark and Zucker 2023). Upon returning to headquarters, these staff bring with them newly acquired insights and may act as effective advocates for rethinking the organization's modus operandi.

Finally, member-states play a central role in guiding the IMF – and other multilateral organizations – toward incorporating climate considerations in their activities. Through their representation in the Executive Board of the IMF, member-states have pushed the organization to reinterpret and extend its mandate in line with evolving scientific views and political consensus (Arias, Clark, and Kaya 2025; Kentikelenis and Seabrooke 2025). After all, it was states who were behind committing to climate goals through the Paris Agreement and related climate targets, reflecting growing domestic political support for climate action (Heinzel et al. 2025).

A full explanation of how these forces – senior leadership, bureaucratic dynamics, and state influence – interact to shape the IMF's climate engagement described in this volume is not attempted here, but it is likely that no single factor dominates. Rather, our ambition has been to set the empirical stage for even beginning to ask these questions. With this foundation in place, subsequent work can now take on this task.

5.2 The Internal Contradictions and Blind Spots of IMF Operations

The IMF's climate-related initiatives reveal significant internal contradictions, particularly between its commitments to climate action and its adherence to excessively restrictive fiscal policies and market-based solutions that can only form a partial response to climate problems (Stern 2008; Stern and Stiglitz 2023). To be sure, some of the reforms promoted by the IMF can have a meaningful impact in fostering climate change adaptation and mitigation, but others may hamper those goals by limiting the policy space of countries to develop and implement comprehensive climate strategies that are also sensitive to just transition considerations.

In this context, the potential harm caused by austerity requires attention, as it can impede the introduction of green transition policies through multiple pathways. Most obviously, austerity measures limit available public financing to scale up investment in climate change adaptation and mitigation strategies. Such investment is unlikely to be pursued in countries that need to decide between cuts to politically sensitive areas of public spending (e.g., education and health) versus climate policies that only have a pay-off in the medium- to long-run. Similarly, austerity may stunt the introduction of any green industrial policies. This is an umbrella term covering multifaceted policies, including subsidies for the introduction of greener production processes, incentives for research and development, tax relief for green investments by the private sector, and redistributive measures to minimize social dislocations stemming from decarbonization initiatives (Allan and Nahm 2024; Hochstetler 2020; Lebdioui 2024; Meckling 2021; Nahm 2021). In all cases, these policies require active state intervention to appropriately and effectively channel green investments. However, notwithstanding IMF research documenting the significance of such policies to complement market-driven approaches (Prasad et al. 2022), IMF-mandated budget cuts and market-oriented reforms have a track record of undermining the state capacity necessary to effectively see this mission through (Reinsberg et al. 2019).

Austerity also has more insidious effects on the likelihood of the green transition. A long-established strand of academic scholarship has shown that the economic performance of countries implementing IMF programs tends to suffer (Dreher 2006). The reasons for this are many and go beyond the observation that countries that resort to IMF programs are generally already going through crisis: ineffective or overambitious IMF-designed policies, inappropriate pacing of reforms, and poor analytical frameworks (e.g., in estimating the fiscal multipliers that inform the design of austerity measures) can all contribute to deepening and prolonging pre-existing economic troubles (see Kentikelenis and Stubbs 2023 for a review of these issues).

Depressed economic activity has climate-damaging follow-on implications. Households find it unaffordable to shift to greener consumption patterns, with low-income ones being particularly hard-pressed. Private firms struggle to invest in green technologies and adaptation measures, thereby increasing their vulnerability to climate shocks and forestalling reductions in their emissions. And the public sector is further starved of revenue sources. In line with the goals set out in the Paris Agreement and the Nationally Determined Contributions that spell out countries' climate plans, domestic resource mobilization is essential for financing green transformation policies, provided they are pursued through progressive means rather than regressive policies like carbon taxes. But austerity-induced economic downturns further reduce tax revenues.

Further, the IMF continues to exhibit several blind spots. First, its debt sustainability analyses rarely incorporate long-term climate adaptation costs. Many climate-vulnerable countries face escalating public debt burdens, yet the IMF's fiscal recommendations do not systematically account for the macroeconomic benefits of proactive climate adaptation investments. This omission risks reinforcing a cycle in which climate shocks exacerbate economic instability, leading to repeated IMF interventions that prioritize short-term fiscal targets over long-term resilience.

Second, the IMF's approach underestimates the broader geopolitical and economic shifts associated with the global energy transition. Many developing countries rely on fossil fuel exports for fiscal revenue, yet IMF policy advice does not always address the fiscal adjustments required in the face of declining global demand for carbon-intensive commodities. The risk of stranded assets and revenue losses from carbon border taxes in major trading partners remains underexplored in IMF reports.

Finally, the IMF's climate policies do not fully engage with the just transition agenda. While there is recognition of the social impacts of subsidy reforms and energy price adjustments, policy responses often focus on narrow compensatory measures rather than structural reforms to create sustainable, equitable green economies. The absence of a clear framework for integrating labor market transitions into IMF-supported reforms limits the institution's ability to support socially inclusive climate policies.

5.3 The IMF in Its Broader Context

The IMF's ongoing transformation toward incorporating climate objectives in its operations has wider significance beyond its headquarters in Washington

DC. This is because of the role of the IMF as a focal point in global economic governance, thus setting the tone for other international financial institutions to emulate or even distinguish themselves from. For example, the IMF's early climate drive took the World Bank by surprise, which then had to innovate by introducing its own diagnostic tools, in direct competition to those provided by the IMF (Kentikelenis, Stubbs, and Reinsberg 2022). Thus, whether due to cooperative relations, competitive dynamics, or organizational imitation, what happens at the IMF has much broader implications.

Insights from the IMF's greening of its operations most readily apply to the global development banks, like the World Bank and European Investment Bank, and regional development banks, such as the Asian Development Bank, Inter-American Development Bank, and European Bank for Reconstruction and Development, all of which are facing pressure to align their operations with the Paris Agreement (Bazbauers 2022; Kentikelenis, Stubbs, and Reinsberg 2022; Michaelowa et al. 2020; Reinsberg et al. 2020; Xie, Scholtens, and Homroy 2023). As with the IMF, the depth of their climate transformation has come into question. For example, Mertens and Thiemann (2023) describe how climate commitments by the European Investment Bank remain subordinate to market logics and fiscal orthodoxy – a finding consistent with our own vis-à-vis the IMF – as the organization manages the trade-off between the high-risk investment needed for the green transition with the imperative of maintaining triple-A credit ratings. This literature suggests that the IMF is not an outlier in its incomplete mainstreaming of the climate agenda but part of a wider family of multilateral financial institutions navigating similar tensions between maintaining core functions and responding to global climate commitments.

Zooming out from international financial institutions, the case of the IMF offers important insights for our understanding of how multilateral organizations in general respond to climate change. We have observed how the Fund's engagement with climate policy represents both progress and constraint. New lending instruments and enhanced surveillance have brought climate considerations into the Fund's operations, but deeper structural changes remain elusive. The continued emphasis on fiscal consolidation, market-based reforms, and private sector-led transitions limits the IMF's ability to fully support the green transition in a manner that aligns with global goals. Indeed, in a global context, these policies appear anachronistic: at the same time as high-income countries have been introducing large-scale green transition policy programs based on active industrial policies and extensive state intervention, the IMF is still promoting an approach that is unlikely to yield more than marginal change in green transition trajectories. Ultimately, tensions between undergoing organizational transformation and maintaining long-established practices are likely present in all major multilateral institutions that have to constantly defend any novel areas of activity with reference to their mandates.

The road ahead is also fraught with uncertainty. At the time of writing (May 2025), the IMF is projecting a future of weaker growth, higher debt burdens, and greater political economy constraints (IMF 2025b). Perhaps the greatest threat to the ongoing efforts to integrate climate considerations into the work of the IMF is the climate-skeptic position of the new US administration, the Fund's largest shareholder. US Treasury Secretary Scott Bessent sharply criticized the IMF in a recent speech for what he called "mission creep ... [and] devoting disproportionate time and resources to work on climate change, gender, and social issues" (US Department of the Treasury 2025). While IMF Managing Director Kristalina Georgieva responded by reasserting the organization's mandate to address balance of payments risks that could arise from a rapidly advancing climate crisis (IMF 2025a), the changing political winds may lead climate issues to be sidelined from the Fund's agenda, especially if the United States and other climate-skeptic countries seek to actively block IMF engagement in such issues. To be sure, the IMF's major organizational innovation – the new Resilience and Sustainability Facility – is not at risk, as the United States is not a contributor to its resource pool. Even so, the United States is the only country with veto power within the IMF and obstructionism can take many forms. This leaves the IMF with difficult choices for how to placate its largest shareholder while not betraying its promises to help countries put in place economic policies that will help meet climate objectives.

Appendix

List of IMF Lending Programs Approved between January 2020 and December 2024

Country	Program Type	RSF-Linked	Date of Approval	Date of Expiration	Loan Size (Thousands of SDRs)	Climate Vulnerability
Afghanistan	ECF		06-Nov-20	06-Dec-22	259,040	High
Argentina	EFF		25-Mar-22	31-Dec-24	31,914,000	Medium
Armenia	SBA		12-Dec-22	11-Dec-25	128,800	Low
Bangladesh	ECF-EFF	Yes	30-Jan-23	29-Jul-26	2,468,460	High
Bangladesh	RSF	ECF-EFF	30-Jan-23	29-Jul-26	1,000,000	High
Barbados	EFF	Yes	07-Dec-22	06-Dec-25	85,050	Low
Barbados	RSF	EFF	07-Dec-22	06-Dec-25	141,750	Low
Benin	ECF-EFF	Yes	08-Jul-22	07-Jan-26	484,058	High
Benin	RSF	ECF-EFF	14-Dec-23	07-Jan-26	148,560	High
Burkina Faso	ECF		21-Sep-23	20-Sep-27	228,760	High
Burundi	ECF		17-Jul-23	16-Sep-26	200,200	High
Cabo Verde	ECF	Yes	15-Jun-22	14-Jun-25	45,030	Medium
Cabo Verde	RSF	ECF	11-Dec-23	14-Jun-25	23,700	Medium
Cameroon	ECF-EFF	Yes	29-Jul-21	28-Jul-25	593,400	High
Cameroon	RSF	ECF-EFF	29-Jan-24	28-Jul-25	138,000	High
Central African Rep.	ECF		27-Apr-23	26-Jun-26	147,480	High

(cont.)

Country	Program Type	RSF-Linked	Date of Approval	Date of Expiration	Loan Size (Thousands of SDRs)	Climate Vulnerability
Chad	ECF		10-Dec-21	21-Jun-24	392,560	High
Chile	FCL		29-May-20	19-May-22	17,443,000	Low
Chile	SLL		20-May-22	22-Aug-28	2,529,000	Low
Chile	FCL		29-Aug-22	26-Aug-24	13,954,000	Low
Chile	FCL		27-Aug-24	26-Aug-26	10,465,800	Low
Colombia	FCL		01-May-20	28-Apr-22	12,267,000	Medium
Colombia	FCL		29-Apr-22	25-Apr-24	7,155,700	Medium
Colombia	FCL		26-Apr-24	25-Apr-26	6,133,500	Medium
Comoros	ECF		01-Jun-23	31-May-27	32,040	High
Congo, Dem. Rep.	ECF		15-Jul-21	09-Jul-24	1,066,000	High
Congo, Rep.	ECF		21-Jan-22	20-Jan-25	324,000	High
Costa Rica	EFF	Yes	01-Mar-21	18-Jun-24	1,237,490	Medium
Costa Rica	RSF	EFF	14-Nov-22	31-Jul-24	554,100	Medium
Cote d'Ivoire	ECF-EFF	Yes	24-May-23	23-Sep-26	2,601,600	High
Cote d'Ivoire	RSF	ECF-EFF	15-Mar-24	23-Sep-26	975,600	High

Country	Type		Start	End	Amount	Risk
Ecuador	EFF		30-Sep-20	16-Dec-22	4,615,000	Medium
Ecuador	EFF		31-May-24	30-May-28	3,000,000	Medium
Egypt	SBA		26-Jun-20	25-Jun-21	3,763,640	Medium
Egypt	EFF		16-Dec-22	15-Oct-26	6,111,690	Medium
Ethiopia	ECF		29-Jul-24	28-Jul-28	2,555,950	High
Gabon	EFF		28-Jul-21	27-Jul-24	388,800	High
Gambia	ECF		23-Mar-20	14-Jun-23	70,550	High
Gambia	ECF		12-Jan-24	11-Jan-27	74,640	High
Georgia	SBA		15-Jun-22	14-Jun-25	210,400	Low
Ghana	ECF		17-May-23	16-May-26	2,241,900	Medium
Guinea-Bissau	ECF		30-Jan-23	29-Jan-26	39,760	High
Honduras	ECF-EFF		21-Sep-23	20-Sep-26	624,500	High
Jamaica	RSF	PLL	01-Mar-23	04-Sep-24	574,350	Medium
Jamaica	PLL	Yes	01-Mar-23	28-Feb-25	727,510	Medium
Jordan	EFF		25-Mar-20	16-Dec-23	1,145,954	Medium
Jordan	EFF		10-Jan-24	09-Jan-28	926,370	Medium
Kenya	ECF-EFF	Yes	02-Apr-21	01-Apr-25	2,714,000	High
Kenya	RSF	ECF-EFF	17-Jul-23	01-Apr-25	407,100	High
Kosovo	RSF	SBA	25-May-23	24-May-25	61,950	Medium
Kosovo	SBA	Yes	25-May-23	24-May-25	80,122	Medium
Liberia	ECF		25-Sep-24	24-Jan-28	155,000	High

(cont.)

Country	Program Type	RSF-Linked	Date of Approval	Date of Expiration	Loan Size (Thousands of SDRs)	Climate Vulnerability
Madagascar	ECF		29-Mar-21	07-Jun-24	219,960	High
Madagascar	ECF	Yes	21-Jun-24	20-Jun-27	256,620	High
Madagascar	RSF	ECF	21-Jun-24	20-Jun-27	244,400	High
Malawi	ECF		15-Nov-23	14-Nov-27	131,860	High
Mauritania	ECF-EFF	Yes	25-Jan-23	24-Jul-26	64,400	High
Mauritania	RSF	ECF-EFF	19-Dec-23	24-Jul-26	193,200	High
Mexico	FCL		19-Nov-21	14-Nov-23	35,650,800	Medium
Mexico	FCL		15-Nov-23	14-Nov-25	26,738,100	Medium
Moldova	ECF-EFF	Yes	20-Dec-21	19-Oct-25	594,263	Medium
Moldova	RSF	ECF-EFF	06-Dec-23	19-Oct-25	129,375	Medium
Morocco	FCL	Yes	03-Apr-23	02-Apr-25	3,726,200	Medium
Morocco	RSF	FCL	28-Sep-23	02-Apr-25	1,000,000	Medium
Mozambique	ECF		09-May-22	08-May-25	340,800	High
Nepal	ECF		12-Jan-22	11-Jan-26	282,420	Medium
Niger	ECF	Yes	08-Dec-21	07-Dec-25	197,400	High
Niger	RSF	ECF	05-Jul-23	07-Dec-25	98,700	High
North Macedonia	PLL		21-Nov-22	20-Nov-24	406,870	Low

Country	Facility	Date		Approval Date	Expiration Date	Amount	Risk
Pakistan	SBA			12-Jul-23	29-Apr-24	2,250,000	High
Pakistan	EFF			25-Sep-24	24-Oct-27	5,320,000	High
Panama	PLL		Yes	19-Jan-21	18-Jan-23	1,884,000	Medium
Papua New Guinea	ECF-EFF			22-Mar-23	21-Dec-26	684,320	High
Papua New Guinea	RSF	ECF-EFF		11-Dec-24	21-Dec-26	197,400	High
Paraguay	RSF	PCI		19-Dec-23	20-Nov-25	302,100	Medium
Peru	FCL			28-May-20	26-May-22	8,007,000	Medium
Peru	FCL			27-May-22	26-May-24	4,003,500	Medium
Rwanda	RSF	PCI-SCF		12-Dec-22	17-Dec-24	240,300	High
Rwanda	SCF		Yes	14-Dec-23	17-Dec-24	200,250	High
Sao Tome & Principe	ECF			19-Dec-24	18-Apr-28	18,500	High
Senegal	SBA-SCF			07-Jun-21	10-Jan-23	582,440	High
Senegal	RSF	ECF-EFF		26-Jun-23	25-Jun-26	242,700	High
Senegal	ECF-EFF		Yes	26-Jun-23	25-Jun-26	1,132,600	High
Serbia	SBA			19-Dec-22	08-Dec-24	1,898,920	Medium
Seychelles	EFF			29-Jul-21	30-May-23	74,000	Medium
Seychelles	RSF	EFF		31-May-23	30-May-26	34,350	Medium
Seychelles	EFF		Yes	31-May-23	30-May-26	42,365	Medium
Sierra Leone	ECF			31-Oct-24	30-Dec-27	186,663	High

(cont.)

Country	Program Type	RSF-Linked	Date of Approval	Date of Expiration	Loan Size (Thousands of SDRs)	Climate Vulnerability
Somalia	ECF-EFF		25-Mar-20	15-Dec-23	292,430	High
Somalia	ECF		19-Dec-23	18-Dec-26	75,000	High
Sri Lanka	EFF		20-Mar-23	19-Mar-27	2,286,000	Medium
Sudan	ECF		29-Jun-21	28-Dec-22	1,733,051	High
Suriname	EFF		22-Dec-21	31-Mar-25	430,700	Medium
Tanzania	ECF	Yes	18-Jul-22	17-May-26	795,580	High
Tanzania	RSF	ECF	20-Jun-24	17-May-26	596,700	High
Togo	ECF		01-Mar-24	31-Aug-27	293,600	High
Uganda	ECF		28-Jun-21	27-Jun-24	722,000	High
Ukraine	SBA		09-Jun-20	02-Mar-22	3,600,000	Medium
Ukraine	EFF		31-Mar-23	30-Mar-27	11,608,250	Medium
Zambia	ECF		31-Aug-22	30-Oct-25	1,271,660	High

Notes: SDRs stands for Special Drawing Rights. Program types are Extended Credit Facility (ECF), Extended Fund Facility (EFF), Flexible Credit Line (FCL), Policy Coordination Instrument (PCI), Precautionary and Liquidity Line (PLL), Resilience and Sustainability Facility (RSF), Short-Term Liquidity Line (SLL), Stand-By Arrangement (SBA), and Stand-By Credit Facility (SCF). We do not report PCI programs as they do not have credit attached. Climate vulnerability score is based on ND-GAIN index ranking (1–62 is low, 63–124 is medium, and 125–187 is high). Kosovo ND-GAIN data was missing so was scored 'Medium' based on our assessment.

References

ActionAid. 2021. *The Public versus Austerity: Why Public Sector Wage Bill Constraints Must End*. London: ActionAid.

ActionAid International. 2023. "The Vicious Cycle: Connections between the Debt Crisis and Climate Crisis." https://actionaid.org/publications/2023/vicious-cycle (March 29, 2024).

Aftab, Hassan. 2022. "Heat Waves and the Ticking Climate Bomb." *The Express Tribune*. https://tribune.com.pk/story/2361084/heat-waves-and-the-ticking-climate-bomb (July 5).

Allan, Bentley B., and Jonas Nahm. 2024. "Strategies of Green Industrial Policy: How States Position Firms in Global Supply Chains." *American Political Science Review* 119(1): 420–34. https://doi.org/10.1017/S0003055424000364.

Arias, Sabrina B., Richard Clark, and Ayse Kaya. 2025. "Power by Proxy: Participation as a Resource in Global Governance." *The Review of International Organizations*. https://doi.org/10.1007/s11558-025-09585-4.

Arinaldo, Deon, and Julius Christian Adiatma. 2019. *Indonesia's Coal Dynamics: Toward a Just Energy Transition*. Jakarta: Institute for Essential Services Reform. IESR Report.

Bazbauers, Adrian Robert. 2022. "Sustainable, Green, and Climate-Resilient Cities: An Analysis of Multilateral Development Banks." *Climate and Development* 14(8): 689–704. https://doi.org/10.1080/17565529.2021.1974331.

Beiser-McGrath, Liam F. 2022. "COVID-19 Led to a Decline in Climate and Environmental Concern: Evidence from UK Panel Data." *Climatic Change* 174(3–4): 31. https://doi.org/10.1007/s10584-022-03449-1.

Beiser-McGrath, Liam F., and Thomas Bernauer. 2019. "Could Revenue Recycling Make Effective Carbon Taxation Politically Feasible?" *Science Advances* 5(9): eaax3323. https://doi.org/10.1126/sciadv.aax3323.

Bettarelli, Luca, Davide Furceri, Pietro Pizzuto, and Nadia Shakoor. 2023. "Environmental Policies and Innovation in Renewable Energy." *IMF Working Paper* 180. www.imf.org/en/Publications/WP/Issues/2023/09/01/Environmental-Policies-and-Innovation-in-Renewable-Energy-538759 (September 27).

Bigger, Patrick, and Sophie Webber. 2021. "Green Structural Adjustment in the World Bank's Resilient City." *Annals of the American Association of Geographers* 111(1): 36–51. https://doi.org/10.1080/24694452.2020.1749023.

Black, Simon, Antung A. Liu, Ian Parry, and Nate Vernon. 2023. *IMF Fossil Fuel Subsidies Data: 2023 Update*. Washington, DC: International Monetary Fund. IMF Working Paper.

Blanchard, Olivier J, and Daniel Leigh. 2014. "Learning about Fiscal Multipliers from Growth Forecast Errors." *IMF Economic Review* 62(2): 179–212. https://doi.org/10.1057/imfer.2014.17.

Blyth, Mark. 2013. *Austerity: The History of a Dangerous Idea*. Oxford: Oxford University Press.

Bocanegra, Nelson. 2022. "Colombia to 'radically Diversify' Exports, Says New Finance Minister." *Reuters*. www.reuters.com/world/americas/colombia-radically-diversify-exports-says-new-finance-minister-2022-08-11/ (September 26, 2023).

Bradlow, Benjamin H., and Alexandros Kentikelenis. 2024. "Globalizing Green Industrial Policy through Technology Transfers." *Nature Sustainability* 7(1): 685–87. https://doi.org/10.1038/s41893-024-01336-4.

Braithwaite, David, and Ivetta Gerasimchuk. 2019. *Beyond Fossil Fuels: Indonesia's Fiscal Transition*. Geneva: Global Subsidies Initiative. GSI Report.

Breen, Michael, and Elliott Doak. 2023. "The IMF as a Global Monitor: Surveillance, Information, and Financial Markets." *Review of International Political Economy* 30(1): 307–31. https://doi.org/10.1080/09692290.2021.2004441.

Broome, André, and Leonard Seabrooke. 2015. "Shaping Policy Curves: Cognitive Authority in Transnational Capacity Building." *Public Administration* 93(4): 956–72. https://doi.org/10.1111/padm.12179.

Burton, Jesse, Tara Caetano, Alison Hughes et al. 2016. *The Impact of Stranding Power Sector Assets in South Africa: Using a Linked Model to Understand Economy-Wide Implications*. Cape Town: Energy Research Centre. Migration Action Plans & Scenarios.

Chateau, Jean, Florence Jaumotte, and Gregor Schwerhoff. 2022. *Economic and Environmental Benefits from International Cooperation on Climate Policies*. Washington, DC: International Monetary Fund. IMF Departmental Paper.

Chowdhury, Anis, and Jomo Kwame Sundaram. 2023. "Chronicles of Debt Crises Foretold." *Development & Change* 54(5): 994–1030. https://doi.org/10.1111/dech.12786.

Clark, Richard, and Noah Zucker. 2023. "Climate Cascades: IOs and the Prioritization of Climate Action." *American Journal of Political Science* 68(4): 1299–314. https://doi.org/10.1111/ajps.12793.

Clift, Ben. 2024. "Global Economic Governance and Environmental Crisis: The Widening Repertoire of IMF Economic Ideas and Limits of Its Climate

Policy Advocacy." *Climate Policy* : 1–15. https://doi.org/10.1080/14693062.2024.2441226.

Clifton, Judith, Daniel Díaz Fuentes, and David J. Howarth, eds. 2021. *Regional Development Banks in the World Economy.* Oxford: Oxford University Press.

Climate Transparency. 2021. *South Africa: Climate Transparency Report 2021.*

Cohen, Michael, and Paul Burkhardt. 2022. "Why Blackouts Are Still Crippling South Africa." *Bloomberg.* www.bloomberg.com/news/articles/2022-07-26/why-blackouts-are-still-crippling-south-africa-quicktake-l6243fsw (August 30).

Copelovitch, Mark, and Stephanie Rickard. 2021. "Partisan Technocrats: How Leaders Matter in International Organizations." *Global Studies Quarterly* 1(3): ksab021. https://doi.org/10.1093/isagsq/ksab021.

Dellmuth, Lisa Maria, and Maria-Therese Gustafsson. 2021. "Global Adaptation Governance: How Intergovernmental Organizations Mainstream Climate Change Adaptation." *Climate Policy* 21(7): 868–83. https://doi.org/10.1080/14693062.2021.1927661.

Dellmuth, Lisa Maria, Maria-Therese Gustafsson, and Ece Kural. 2020. "Global Adaptation Governance: Explaining the Governance Responses of International Organizations to New Issue Linkages." *Environmental Science & Policy* 114: 204–15. https://doi.org/10.1016/j.envsci.2020.07.027.

Dellmuth, Lisa Maria, Maria-Therese Gustafsson, Niklas Bremberg, and Malin Mobjörk. 2018. "Intergovernmental Organizations and Climate Security: Advancing the Research Agenda." *WIREs Climate Change* 9(1): e496. https://doi.org/10.1002/wcc.496.

Deloitte. 2022. "Wide-Ranging Tax Reform Enacted." *tax@hand.* www.taxathand.com/article/27284/Colombia/2022/Wide-ranging-tax-reform-enacted (August 15, 2023).

Dörfler, Thomas, and Mirko Heinzel. 2023. "Greening Global Governance: INGO Secretariats and Environmental Mainstreaming of IOs, 1950 to 2017." *The Review of International Organizations* 18(1): 117–43. https://doi.org/10.1007/s11558-022-09462-4.

Dreher, Axel. 2006. "IMF and Economic Growth: The Effects of Programs, Loans, and Compliance with Conditionality." *World Development* 34(5): 769–88. https://doi.org/10.1016/j.worlddev.2005.11.002.

Edwards, Martin S., and Stephanie Senger. 2015. "Listening to Advice: Assessing the External Impact of IMF Article IV Consultations of the United States, 2010–2011." *International Studies Perspectives* 16(3): 312–26. https://doi.org/10.1111/insp.12059.

Ellis-Petersen, Hannah, and Shah Meer Baloch. 2022. "'We Are Living in Hell': Pakistan and India Suffer Extreme Spring Heatwaves." *The Guardian.* www

.theguardian.com/world/2022/may/02/pakistan-india-heatwaves-water-elec
tricity-shortages (October 9).

Ethirajan, Anbarasan. 2023. "Pakistan Economy: Price Spikes, Political Woes Hit Millions." *BBC News.* www.bbc.com/news/world-asia-67494013 (February 10, 2025).

Fellesson, Måns. 2017. *Research Capacity in the New Global Development Agenda: Mobility, Collaboration and Scientific Production among PhD Graduates Supported by Swedish Development Aid in Africa.* Stockholm: Expertgruppen för Biståndsanalys.

Ferguson, Thomas, and Servaas Storm. 2023. "Central Banks Raising Interest Rates Makes It Harder to Fight the Climate Crisis." *The Guardian.* www.theguardian.com/commentisfree/2023/may/06/central-banks-interest-rate-hike-climate-crisis (August 21).

Fischer, Andrew M., and Servaas Storm. 2023. "The Return of Debt Crisis in Developing Countries: Shifting or Maintaining Dominant Development Paradigms?" *Development & Change* 54(5): 954–93. https://doi.org/10.1111/dech.12800.

Fratzscher, Marcel, and Julien Reynaud. 2011. "IMF Surveillance and Financial Markets – – A Political Economy Analysis." *European Journal of Political Economy* 27(3): 405–22. https://doi.org/10.1016/j.ejpoleco.2011.01.002.

Frenkel, Michael, Jan-Christoph Rülke, and Lilli Zimmermann. 2013. "Do Private Sector Forecasters Chase after IMF or OECD Forecasts?" *Journal of Macroeconomics* 37: 217–29. https://doi.org/10.1016/j.jmacro.2013.03.002.

Fresnillo, Iolanda. 2024. "Debt Justice in 2024: Challenges and Prospects in a Full-Blown Debt Crisis." *Eurodad.* www.eurodad.org/debt_justice_in_2024_challenges_and_prospects_in_a_full_blown_debt_crisis (March 8, 2024).

Fundación Ambiente y Recursos Naturales. 2019. *Fossil Fuel Subsidies in Argentina 2018–2019.* Buenos Aires: Fundación Ambiente y Recursos Naturales.

Fundación Ambiente y Recursos Naturales. 2021. *La Transición Energética En La Argentina y Los Subsidios a Los Fósiles.* Buenos Aires: Fundación Ambiente y Recursos Naturales.

G20. 2021. "Rome Leaders' Declaration." www.governo.it/sites/governo.it/files/G20ROMELEADERSDECLARATION_0.pdf (January 31, 2025).

Gabor, Daniela. 2021. "The Wall Street Consensus." *Development and Change* 52(3): 429–59. https://doi.org/10.1111/dech.12645.

Gabor, Daniela, and Benjamin Braun. 2025. "Green Macrofinancial Regimes." *Review of International Political Economy* 32(3): 542–68. https://doi.org/10.1080/09692290.2025.2453504.

Gallagher, Kevin, Cyrus Rustomjee, and Andrea Arevalo. 2024. "Evolution of IMF Engagement on Climate Change." *IEO Background Paper*: BP/24–01/06.

Gallagher, Kevin P., Luma Ramos, Corinne Stephenson, and Irene Monasterolo. 2021. "Climate Change and IMF Surveillance: The Need for Ambition." *GEGI Policy Brief* 014. Boston, MA: Global Development Policy Center.

Georgieva, Kristalina. 2021. "Remarks by IMF Managing Director on Global Policies and Climate Change." *IMF Speech*. www.imf.org/en/News/Articles/2021/07/11/sp071121-md-on-global-policies-and-climate-change (July 23).

Government of Colombia. 2023. *Marco Fiscal de Mediano Plazo 2023: Estrategia Sostenible para La Transformación Social y Económica de Colombia*. Bogotá: Government of Colombia.

Government of Indonesia. 2022. *Enhanced Nationally Determined Contribution*. Jakarta: Government of Indonesia.

Government of Kenya. 2020. *Kenya's Updated Nationally Determined Contributions (NDC)*. Nairobi: Government of Kenya.

Government of South Africa. 2020. *The South African Economic Reconstruction and Recovery Plan*. Pretoria: Government of South Africa.

Government of South Africa. 2021. *First Nationally Determined Contribution under the Paris Agreement*. Cape Town: Government of South Africa.

Hagan, Sean. 2023. "More DSA Transparency, Please." *Financial Times*. www.ft.com/content/2d5f3af5-c34d-406c-b9f0-056efc83cc51 (March 8, 2024).

Hall, Nina. 2015. "Money or Mandate? Why International Organizations Engage with the Climate Change Regime." *Global Environmental Politics* 15(2): 79–97. https://doi.org/10.1162/GLEP_a_00299.

Heinzel, Mirko, Andreas Kern, Saliha Metinsoy, and Bernhard Reinsberg. 2025. "Public Support for Green, Inclusive, and Resilient Growth Conditionality in International Monetary Fund Bailouts." *International Studies Quarterly* 69(2): sqaf018. https://doi.org/10.1093/isq/sqaf018.

Heinzel, Mirko, Catherine Weaver, and Samantha Jorgensen. 2024. "Bureaucratic Representation and Gender Mainstreaming in International Organizations: Evidence from the World Bank." *American Political Science Review* 119 (1):332–48. https://doi.org/10.1017/S0003055424000376.

Hochstetler, Kathryn. 2020. *Political Economies of Energy Transition: Wind and Solar Power in Brazil and South Africa*. Cambridge: Cambridge University Press.

Hosan, Shahadat, Kanchan Kumar Sen, Md. Matiar Rahman et al. 2023. "Evaluating the Mediating Role of Energy Subsidies on Social Well-Being and Energy Poverty Alleviation in Bangladesh." *Energy Research & Social Science* 100: 103088. https://doi.org/10.1016/j.erss.2023.103088.

Humphrey, Chris. 2022. *Financing the Future: Multilateral Development Banks in the Changing World Order of the 21st Century.* Oxford: Oxford University Press. https://doi.org/10.1093/oso/9780192871503.001.0001.

Huxham, Matthew, Muhammed Anwar, and David Nelson. 2019. *Understanding the Impact of a Low Carbon Transition on South Africa.* London: Climate Policy Initiative. CPI Energy Finance Report.

IEA. 2024a. "Argentina." www.iea.org/countries/argentina (January 29, 2025).

IEA. 2024b. "Colombia." www.iea.org/countries/colombia (December 18).

IEA. 2024c. "Indonesia." www.iea.org/countries/indonesia (December 12).

IEA. 2024d. "Pakistan." www.iea.org/countries/pakistan (January 29, 2025).

IEA. 2024e. "South Africa." www.iea.org/countries/south-africa (December 16).

Ilyina, Anna, Ceyla Pazarbasioglu, and Michele Ruta. 2024. "Industrial Policy Is Back but the Bar to Get It Right Is High." *IMF.* www.imf.org/en/Blogs/Articles/2024/04/12/industrial-policy-is-back-but-the-bar-to-get-it-right-is-high (August 31).

IMF. 2019a. *2018 Review of Program Design and Conditionality.* Washington, DC: International Monetary Fund. IMF Policy Paper.

IMF. 2019b. *A Strategy for IMF Engagement on Social Spending.* Washington, DC: International Monetary Fund. IMF Policy Paper.

IMF. 2019c. *Building Resilience in Developing Countries Vulnerable to Large Natural Disasters.* Washington, DC: International Monetary Fund. IMF Policy Paper.

IMF. 2020a. *Articles of Agreement.* Washington, DC: International Monetary Fund.

IMF. 2020b. "Transcript of International Monetary Fund Managing Director Kristalina Georgieva's Opening Press Conference, 2020 Spring Meetings." *IMF Transcript.* www.imf.org/en/News/Articles/2020/04/15/tr041520-transcript-of-imf-md-kristalina-georgieva-opening-press-conference-2020-spring-meetings (December 19).

IMF. 2020c. *World Economic Outlook, October 2020: A Long and Difficult Ascent.* Washington, DC: International Monetary Fund.

IMF. 2021a. *2021 Comprehensive Surveillance Review – Background Paper on Integrating Climate Change into Article IV Consultations.* Washington, DC: International Monetary Fund. IMF Policy Paper.

IMF. 2021b. *2021 Comprehensive Surveillance Review – Overview Paper.* Washington, DC: International Monetary Fund. IMF Policy Paper.

IMF. 2021c. *2021 Financial Sector Assessment Program Review – Towards a More Stable and Sustainable Financial System.* Washington, DC: International Monetary Fund. IMF Policy Paper.

IMF. 2021d. *IMF Strategy to Help Members Address Climate Change Related Policy Challenges: Priorities, Modes of Delivery, and Budget Implications.* Washington, DC: International Monetary Fund. IMF Policy Paper.

IMF. 2021e. *Indonesia: Selected Issues.* Washington, DC: International Monetary Fund. IMF Country Report.

IMF. 2021f. *Kenya: Requests for an Extended Arrangement under the Extended Fund Facility and an Arrangement under the Extended Credit Facility.* Washington, DC: International Monetary Fund. IMF Country Report.

IMF. 2021g. *Pakistan: Second, Third, Fourth and Fifth Reviews under the Extended Arrangement under the Extended Fund Facility and Request for Rephasing of Access – Press Release; Staff Report; Staff Supplement, and Statement by the Executive Director for Pakistan.* Washington, DC: International Monetary Fund. IMF Country Report.

IMF. 2022a. "2022 Review of Adequacy of Poverty Reduction and Growth Trust Finances." www.imf.org/en/Publications/Policy-Papers/Issues/2022/04/21/2022-Review-of-Adequacy-of-Poverty-Reduction-and-Growth-Trust-Finances-517091 (March 1, 2024).

IMF. 2022b. *Argentina: Staff Report for 2022 Article IV Consultation and Request for an Extended Arrangement under the Extended Fund Facility.* Washington, DC: International Monetary Fund. IMF Country Report.

IMF. 2022c. *Colombia: Financial System Stability Assessment.* Washington, DC: International Monetary Fund. IMF Country Report.

IMF. 2022d. *Ecuador: Sixth Review under the Extended Arrangement under the Extended Fund Facility and Financing Assurances Review.* Washington, DC: International Monetary Fund. IMF Country Report.

IMF. 2022e. *Gabon: First and Second Reviews of the Extended Arrangement under the Extended Fund Facility, Requests for Waivers for Nonobservance of Performance Criteria, Establishment of Performance Criteria, and Financing Assurances Review.* Washington, DC: International Monetary Fund. IMF Country Report.

IMF. 2022f. *Georgia: First Review under the Stand-By Arrangement and Request for Modifications of Performance Criteria and Structural Benchmarks.* Washington, DC: International Monetary Fund. IMF Country Report.

IMF. 2022g. *Georgia: Request for a Stand-By Arrangement.* Washington, DC: International Monetary Fund. IMF Country Report.

IMF. 2022h. *Guidance Note for Surveillance under Article IV Consultations.* Washington, DC: International Monetary Fund. IMF Policy Paper.

IMF. 2022i. *Kenya: Third Reviews under the Extended Arrangement under the Extended Fund Facility and under the Arrangement under the Extended*

Credit Facility, Requests for Modification of Quantitative Performance Criteria, and Waiver of Applicability for Performance Criteria. Washington, DC: International Monetary Fund. IMF Country Report.

IMF. 2022j. *Pakistan: 2021 Article IV Consultation, Sixth Review under the Extended Arrangement under the Extended Fund Facility, and Requests for Waivers of Applicability and Nonobservance of Performance Criteria and Rephasing of Access*. Washington, DC: International Monetary Fund. IMF Country Report.

IMF. 2022k. *Proposal to Establish a Resilience and Sustainability Trust*. Washington, DC: International Monetary Fund. IMF Policy Paper.

IMF. 2022l. *South Africa: 2021 Article IV Consultation*. Washington, DC: International Monetary Fund. IMF Country Report.

IMF. 2022m. *Staff Guidance Note on the Sovereign Risk and Debt Sustainability Framework*. Washington, DC: International Monetary Fund. IMF Policy Paper.

IMF. 2023a. *Chad: First and Second Reviews under the Extended Credit Facility Arrangement, Requests for Waivers of Nonobservance of Performance Criteria and Modification of Performance Criteria*. Washington, DC: International Monetary Fund. IMF Country Report.

IMF. 2023b. *Colombia: 2023 Article IV Consultation*. Washington, DC: International Monetary Fund. IMF Country Report.

IMF. 2023c. *Colombia: Selected Issues*. Washington, DC: International Monetary Fund. IMF Country Report.

IMF. 2023d. *Honduras: 2023 Article IV Consultation and Requests for an Arrangement under the Extended Fund Facility and an Arrangement under the Extended Credit Facility*. Washington, DC: International Monetary Fund. IMF Country Report.

IMF. 2023e. *Indonesia: 2023 Article IV Consultation*. Washington, DC: International Monetary Fund. IMF Country Report.

IMF. 2023f. *Kenya: Fifth Reviews under the Extended Fund Facility and Extended Credit Facility Arrangements and Request for a 20-Month Arrangement under the Resilience and Sustainability Facility, Requests for Extension, Rephasing, and Augmentation of Access, Modification of a Performance Criterion, Waiver of Applicability for Performance Criteria and Waiver of Nonobservance for a Performance Criterion, and Monetary Policy Consultation Clause*. Washington, DC: International Monetary Fund. IMF Country Report.

IMF. 2023g. "*Resilience and Sustainability Facility – Operational Guidance Note.*" Washington, DC: International Monetary Fund. IMF Policy Paper.

IMF. 2023h. *Senegal: First Reviews under the Extended Fund Facility, the Extended Credit Facility, and the Resilience and Sustainability Facility Arrangements, Requests for Modification of the Quantitative Performance Criteria Rephasing of Access.* Washington, DC: International Monetary Fund. IMF Country Report.

IMF. 2023i. *Senegal: Requests for an Extended Arrangement under the Extended Fund Facility, an Arrangement under the Extended Credit Facility, and an Arrangement under the Resilience and Sustainability Facility.* Washington, DC: International Monetary Fund. IMF Country Report.

IMF. 2023j. "World Economic Outlook Database, October." *IMF.* www.imf.org/en/Publications/WEO/weo-database/2023/October.

IMF. 2024a. "Climate Public Investment Management Assessment." https://infrastructuregovern.imf.org/content/PIMA/Home/PimaTool/C-PIMA.html (March 15).

IMF. 2024b. *Democratic Republic of the Congo: 2024 Article IV Consultation, Sixth Review under the Extended Credit Facility Arrangement and Request for Waiver of Nonobservance of Quantitative Performance Criterion, and Financing Assurances Review.* Washington, DC: International Monetary Fund. IMF Country Report.

IMF. 2024c. "DIGNAD: A Toolkit for Macro Policy Assessments of Building Resilience to and Recovery from Natural Disasters in Emerging and Developing Countries." https://climatedata.imf.org/pages/dignad (March 1).

IMF. 2024d. *Ecuador: Request for an Extended Arrangement under the Extended Fund Facility.* Washington, DC: International Monetary Fund. IMF Country Report.

IMF. 2024e. *Jordan: First Review under the Extended Arrangement under the Extended Fund Facility and Request for Modification of Performance Criteria.* Washington, DC: International Monetary Fund. IMF Country Report.

IMF. 2024f. *Kenya: 2023 Article IV Consultation, Sixth Reviews under the Extended Fund Facility and Extended Credit Facility Arrangements, Requests for Augmentations of Access, Modification of Performance Criteria, Waiver of Nonobservance of Performance Criteria, Waiver of Applicability of Performance Criteria, and First Review under the Resilience and Sustainability Facility Arrangement.* Washington, DC: International Monetary Fund. IMF Country Report.

IMF. 2024g. "List of LIC DSAs for PRGT-Eligible Countries: As of October 31, 2024." www.imf.org/external/pubs/ft/dsa/dsalist.pdf (January 30, 2025).

IMF. 2024h. *Republic of Congo: Fourth Review under the Three-Year Arrangement under the Extended Credit Facility, Requests for Modification of Performance Criteria, Waivers of Nonobservance of Performance Criteria, and Financing Assurances Review.* Washington, DC: International Monetary Fund. IMF Country Report.

IMF. 2024i. *Republic of Serbia: Third Review under the Stand-By Arrangement and Request for Modification of Performance Criteria.* Washington, DC: International Monetary Fund. IMF Country Report.

IMF. 2024j. *Suriname: Fourth Review under the Extended Arrangement under the Extended Fund Facility, Requests for Extension of the Arrangement, Augmentation of Access, Modification of Performance Criteria, and Financing Assurances Review.* Washington, DC: International Monetary Fund. IMF Country Report.

IMF. 2024k. *Uganda: Fifth Review under the Extended Credit Facility Arrangement and Request for Modification of Performance Criteria.* Washington, DC: International Monetary Fund. IMF Country Report.

IMF. 2024l. *Zambia: Third Review under the Arrangement under the Extended Credit Facility, Requests for Augmentation of Access, Modifications of the Monetary Policy Consultation Clause and of Quantitative Performance Criteria, and Financing Assurances Review.* Washington, DC: International Monetary Fund. IMF Country Report.

IMF. 2025a. "Press Briefing Transcript: Managing Director's Global Policy Agenda, Spring Meetings 2025." www.imf.org/en/News/Articles/2025/04/24/tr-042425-managing-directors-press-briefing-on-gpa (May 14).

IMF. 2025b. "World Economic Outlook, April 2025: A Critical Juncture amid Policy Shifts." www.imf.org/en/Publications/WEO/Issues/2025/04/22/world-economic-outlook-april-2025 (May 14).

International Crisis Group. 2024. "Sri Lanka's Bailout Blues: Elections in the Aftermath of Economic Collapse, Crisis Group." *International Crisis Group.* www.crisisgroup.org/asia/south-asia/sri-lanka/341-sri-lankas-bailout-blues-elections-aftermath-economic-collapse (November 8).

Ioanes, Ellen. 2023. "Ecuador's Political Instability, Explained." *Vox.* www.vox.com/world-politics/2023/4/30/23705442/ecuador-lasso-political-corruption (November 8, 2024).

Kentikelenis, Alexandros, and Leonard Seabrooke. 2025. *Making Global Norms: Politics versus Science in International Organizations.* Oxford: Oxford University Press.

Kentikelenis, Alexander E., and Sarah L. Babb. 2022. "International Financial Institutions: Forms, Functions, and Controversies." In *The Oxford Handbook of International Political Economy,* eds. Jon C. W. Pevehouse and

Leonard Seabrooke, 351–71. Oxford: Oxford University Press. https://academic.oup.com/edited-volume/35412/chapter/378274496.

Kentikelenis, Alexander, and Thomas Stubbs. 2021a. *Missing Links: How Climate Change Remains Peripheral to IMF Economic Surveillance Activities*. Amsterdam: Recourse. Recourse Report.

Kentikelenis, Alexander, and Thomas Stubbs. 2021b. *Out of the Shadows: Integrating Climate Change into IMF Technical Assistance*. Amsterdam: Recourse. Recourse Report.

Kentikelenis, Alexandros, and Thomas Stubbs. 2023. *A Thousand Cuts: Social Protection in the Age of Austerity*. Oxford: Oxford University Press.

Kentikelenis, Alexandros, and Thomas Stubbs. 2024. "Social Protection and the International Monetary Fund: Promise versus Performance." *Globalization & Health* 20(1): 41. https://doi.org/10.1186/s12992-024-01045-9.

Kentikelenis, Alexander, Thomas Stubbs, and Bernhard Reinsberg. 2022. *The IMF and the Road to a Green and Inclusive Recovery after Covid-19*. Cambridge Centre for Business Research (CBR) Special Report. https://doi.org/10.17863/CAM.86038.

Kling, Gerhard, Ulrich Volz, Victor Murinde, and Sibel Ayas. 2021. "The Impact of Climate Vulnerability on Firms' Cost of Capital and Access to Finance." *World Development* 137: 105131. https://doi.org/10.1016/j.worlddev.2020.105131.

Lagarde, Christine. 2015. "'Lifting the Small Boats,' Speech by IMF Managing Director." *IMF*. www.imf.org/en/News/Articles/2015/09/28/04/53/sp061715 (July 28, 2021).

Lagarde, Christine, and Vitor Gaspar. 2019. "Getting Real on Meeting Paris Climate Change Commitments." *IMF Blog*. www.imf.org/en/Blogs/Articles/2019/05/03/blog-getting-real-on-meeting-paris-climate-change-commitments (September 23, 2022).

Lang, Valentin, Lukas Wellner, and Alexandros Kentikelenis. 2024. "Biased Bureaucrats and the Policies of International Organizations." *American Journal of Political Science*. https://doi.org/10.1111/ajps.12921.

Lebdioui, Amir. 2024. *Survival of the Greenest: Economic Transformation in a Climate-Conscious World*. Cambridge: Cambridge University Press. https://doi.org/10.1017/9781009339414.

Maldonado, Franco, and Kevin P. Gallagher. 2022. "Climate Change and IMF Debt Sustainability Analysis." *Task Force on Climate, Development and the IMF*. www.bu.edu/gdp/files/2022/02/TF_WP_005_FIN.pdf (March 1, 2024).

Mangi, Faseeh. 2024. "Protests in Pakistan after Govt Raises Taxes by 40% under IMF Bailout." *Business Standard India*. www.business-standard.com/

world-news/protests-in-pakistan-after-regime-raise-taxes-by-40-under-imf-bailout-124100800380_1.html (February 10, 2025).

Mazzucato, Mariana, Vera Songwe, Amir Lebdioui et al. 2024. *A Green and Just Planet: 1.5°C Agenda for Governing Global Industrial and Financial Policies in the G20*. Brasilia: Group of Experts to the G20 Taskforce on a Global Mobilization against Climate Change.

Meckling, Jonas. 2021. "Making Industrial Policy Work for Decarbonization." *Global Environmental Politics*, 21(4): 134–47. https://doi.org/10.1162/glep_a_00624.

Mertens, Daniel, and Matthias Thiemann. 2017. "Building a Hidden Investment State? The European Investment Bank, National Development Banks and European Economic Governance." *Journal of European Public Policy* 26(1): 23–43.

Mertens, Daniel, and Matthias Thiemann. 2018. "Market-Based but State-Led: The Role of Public Development Banks in Shaping Market-Based Finance in the European Union." *Competition & Change* 22(2): 184–204. https://doi.org/10.1177/1024529418758479.

Mertens, Daniel, and Matthias Thiemann. 2023. "The European Investment Bank: The EU's Climate Bank?" In *Handbook on European Union Climate Change Policy and Politics*, eds. Tim Rayner, Kacper Szulecki, Andrew J. Jordan, and Sebastian Oberthür. Cheltenham: Edward Elgar, 68–82. www.elgaronline.com/edcollchap-oa/book/9781789906981/book-part-9781789906981-16.xml (May 14, 2025).

Meyer, John W., and Brian Rowan. 1977. "Institutionalized Organizations: Formal Structure as Myth and Ceremony." *American Journal of Sociology* 83(2): 340–63.

Michaelowa, Katharina, Axel Michaelowa, Bernhard Reinsberg, and Igor Shishlov. 2020. "Do Multilateral Development Bank Trust Funds Allocate Climate Finance Efficiently?" *Sustainability* 12(14): 5529. https://doi.org/10.3390/su12145529.

Momani, Bessma. 2006. "Assessing the Utility of, and Measuring Learning from, Canada's IMF Article IV Consultations." *Canadian Journal of Political Science* 39(2): 249–69. https://doi.org/10.1017/S0008423906060124.

de Mooij, Ruud A., Michael Keen, and Ian W. H. Parry. 2012. *Fiscal Policy to Mitigate Climate Change: A Guide to Policymakers*. Washington, DC: Internat. Monetary Fund.

Moulvi, Zain. 2022. "IMF Programme in Pakistan Undermines Renewable Energy Roll-Out." *Bretton Woods Observer*. www.brettonwoodsproject.org/2022/04/imf-programme-in-pakistan-undermines-renewable-energy-roll-out/ (June 1).

Nahm, Jonas. 2021. *Collaborative Advantage: Forging Green Industries in the New Global Economy*. New York: Oxford University Press.

Naqvi, Natalya, Anne Henow, and Ha-Joon Chang. 2018. "Kicking Away the Financial Ladder? German Development Banking under Economic Globalisation." *Review of International Political Economy* 37(4): 1–27. https://doi.org/10.1080/13563469608406251.

Nasruddin, Citra Handayani. 2022. "Gender on the Agenda for Indonesia's Social Protection Policies." www.eastasiaforum.org/2022/11/17/gender-on-the-agenda-for-indonesias-social-protection-policies/ (September 27, 2023).

Nelson, Stephen C. 2014. "Playing Favorites: How Shared Beliefs Shape the IMF's Lending Decisions." *International Organization*, 68(2): 297–328. https://doi.org/10.1017/S0020818313000477.

Nelson, Stephen C. 2017. *The Currency of Confidence: How Economic Beliefs Shape the IMF's Relationship with Its Borrowers*. Ithaca: Cornell University Press.

Ngoma, Hambulo, Patrick Lupiya, Mulako Kabisa, and Faaiqa Hartley. 2021. "Impacts of Climate Change on Agriculture and Household Welfare in Zambia: An Economy-Wide Analysis." *Climatic Change* 167(3–4): 55. https://doi.org/10.1007/s10584-021-03168-z.

Nogrady, Bianca. 2021. "China Launches World's Biggest Carbon Market." *Nature* 595: 637. https://doi.org/10.1038/d41586-021-01989-7.

Notre Dame Global Adaptation Initiative. 2023. "ND-GAIN Country Index Rankings." https://gain.nd.edu/our-work/country-index/rankings/ (May 2).

OEC. 2024. "Indonesia." https://oec.world/en/profile/country/idn (December 12).

Ortiz, Isabel, and Matthew Cummins. 2021. "The Austerity Decade 2010–20." *Social Policy and Society* 20(1): 142–57. https://doi.org/10.1017/S1474746420000433.

Ostry, Jonathan D., Prakash Loungani, and Davide Furceri. 2016. "Neoliberalism: Oversold?" *Finance & Development* 53(2): 38–41.

Parry, Ian, Simon Black, and Nate Vernon. 2021. *Still Not Getting Energy Prices Right: A Global and Country Update of Fossil Fuel Subsidies*. Washington, DC: International Monetary Fund. IMF Working Paper.

Persson, Åsa, and Adis Dzebo. 2019. "Exploring Global and Transnational Governance of Climate Change Adaptation." *International Environmental Agreements: Politics, Law and Economics* 19(4): 357–67. https://doi.org/10.1007/s10784-019-09440-z.

Prasad, Ananthakrishnan, Elena Loukoianova, Alan Xiaochen Feng, and William Oman. 2022. *Mobilizing Private Climate Financing in Emerging Market and Developing Economies*. IMF Staff Climate Note.

https://elibrary.imf.org/openurl?genre=journal&issn=2789-0600&volume=2022&issue=007 (September 2).

Prasetiyo, Andri, Isabella Suarez, Jobit Parapat, and Zakki Amali. 2023. *Ambiguities versus Ambition: A Review of Indonesia's Energy Transition Policy*. Centre for Research on Energy and Clean Air.

Przeworski, Adam, and James Raymond Vreeland. 2000. "The Effect of IMF Programs on Economic Growth." *Journal of Development Economics* 62(2): 385–421. https://doi.org/10.1016/s0304-3878(00)00090-0.

Ramos, Luma, Kevin P. Gallagher, Corinne Stephenson, and Irene Monasterolo. 2022a. "Climate Risk and IMF Surveillance Policy: A Baseline Analysis." *Climate Policy* 22(3): 371–88. https://doi.org/10.1080/14693062.2021.2016363.

Ramos, Luma, Kevin P. Gallagher, William N. Kring, and Franco Maldonado Carlin. 2022b. "The IMF and Climate Change: Tracking the IMF's Engagement and Leadership." *GEGI Working Paper* 056.

Ramos, Luma, Rishikesh Bhandary, Kevin Gallagher, and Rebecca Ray. 2022c. "V20 Debt Review: An Account of Debt in the Vulnerable Group of Twenty." *V20*. www.v-20.org/resources/publications/v20-debt-review (April 1, 2024).

Rehbein, Kristina. 2023. "Understanding Debt Sustainability Analyses." *Friedrich Ebert Stiftung*. https://library.fes.de/pdf-files/bueros/tunesien/20619-20231115.pdf (March 1, 2024).

Reinsberg, Bernhard, Alexander E. Kentikelenis, Thomas H. Stubbs, and Lawrence P. King. 2019. "The World System and the Hollowing Out of State Capacity: How Structural Adjustment Programs Affect Bureaucratic Quality in Developing Countries." *American Journal of Sociology* 124(4): 1222–57. https://doi.org/10.1086/701703.

Reinsberg, Bernhard, Igor Shishlov, Katharina Michaelowa, and Axel Michaelowa. 2020. *Climate Change-Related Trust Funds at the Multilateral Development Banks*. Bonn: Deutsche Gesellschaft für Internationale Zusammenarbeit (GIZ). www.zora.uzh.ch/id/eprint/188309 (May 14, 2025).

Reuters. 2023. "Enel Suspends Colombia Wind Farm Construction after Years of Protests." *Reuters*. www.reuters.com/business/energy/enel-suspends-colombia-wind-farm-construction-after-years-protests-2023-05-24/ (August 21).

Rodrik, D. 2014. "Green Industrial Policy." *Oxford Review of Economic Policy* 30(3): 469–91. https://doi.org/10.1093/oxrep/gru025.

Rumble, Olivia, and Elizabeth Sidiropoulos. 2022. *Exploring the Potential Role of the IMF in Supporting South Africa's Just Transition*. Johannesburg: South African Institute of International Affairs. SAIIA Special Report.

Sanyal, Bikas, and N. V. Varghese. 2006. *Research Capacity of the Higher Education Sector in Developing Countries*. Paris: UNESCO.

Schäfer, Armin. 2006. "A New Form of Governance? Comparing the Open Method of Co-Ordination to Multilateral Surveillance by the IMF and the OECD." *Journal of European Public Policy* 13(1): 70–88. https://doi.org/10.1080/13501760500380742.

Simmons, Beth A., Frank Dobbin, and Geoffrey Garrett, eds. 2008. *The Global Diffusion of Markets and Democracy*. Cambridge: Cambridge University Press.

Simon, Julia. 2023. "Despite Billions to Get off Coal, Why Is Indonesia Still Building New Coal Plants?" *National Public Radio*. www.npr.org/2023/02/05/1152823939/despite-billions-to-get-off-coal-why-is-indonesia-still-building-new-coal-plants (August 25).

Skovgaard, Jakob. 2021. *The Economisation of Climate Change: How the G20, the OECD and the IMF Address Fossil Fuel Subsidies and Climate Finance*. Cambridge: Cambridge University Press.

Stern, Nicholas. 2008. "The Economics of Climate Change." *American Economic Review* 98(2): 1–37. https://doi.org/10.1257/aer.98.2.1.

Stern, Nicholas, and Joseph E. Stiglitz. 2023. "Climate Change and Growth." *Industrial and Corporate Change* 32(2): 277–303. https://doi.org/10.1093/icc/dtad008.

Stubbs, Thomas, and Alexander Kentikelenis. 2023. *IMF Lending and the Road to Green Transition: One Step Forward, One Step Back*. Amsterdam: Recourse. Recourse Report.

Stubbs, Thomas, William Kring, Christina Laskaridis, Alexander Kentikelenis, and Kevin Gallagher. 2021. "Whatever It Takes? The Global Financial Safety Net, Covid-19, and Developing Countries." *World Development* 137: 105171. https://doi.org/10.1016/j.worlddev.2020.105171.

Sward, Jon, Niranjali Amerasinghe, Andrew Bunker, and Jo Walker. 2021. *IMF Surveillance and Climate Change Transition Risks: Reforming IMF Policy Advice to Support a Just Energy Transition*. London: ActionAid.

University of Notre Dame. 2023. "Notre Dame Global Adaptation Initiative (ND-GAIN)." https://gain.nd.edu/our-work/country-index/ (May 26).

US Department of the Treasury. 2025. "Treasury Secretary Scott Bessent Remarks before the Institute of International Finance." https://home.treasury.gov/news/press-releases/sb0094 (May 14).

Volz, Ulrich, and Sara Jane Ahmed. 2020a. "Macrofinancial Risks in Climate Vulnerable Developing Countries and the Role of the IMF." *SOAS*. https://eprints.soas.ac.uk/34348/1/Volz%20Ahmed%202020%20Macrofinancial%

20Risks%20in%20Climate%20Vulnerable%20Developing%20Countries%20and%20the%20Role%20of%20the%20IMF.pdf (December 3, 2021).

Volz, Ulrich, and Sara Jane Ahmed. 2020b. *Macrofinancial Risks in Climate Vulnerable Developing Countries and the Role of the IMF: Towards a Joint V20-IMF Action Agenda*. London: Centre for Sustainable Finance.

Weaver, Catherine, Mirko Heinzel, Samantha Jorgensen, and Joseph Flores. 2022. "Bureaucratic Representation in the IMF and the World Bank." *Global Perspectives* 3(1): 39684. https://doi.org/10.1525/gp.2022.39684.

Wheatley, Jonathan. 2023. "Critics Accuse IMF over Countries' Debt Deadlocks." *Financial Times*. www.ft.com/content/13e77edd-dcb0-4275-b51c-0daf44eec760 (March 8, 2024).

Woolfenden, Tess. 2023a. "The Debt-Fossil Fuel Trap: Why Debt Is a Barrier to Fossil Fuel Phase-out and What We Can Do about It." https://debtjustice.org.uk/wp-content/uploads/2023/08/Debt-Fossil-Fuel-Trap-Report_2023.pdf (March 29, 2024).

Woolfenden, Tess. 2023b. *The Debt-Fossil Fuel Trap: Why Debt Is a Barrier to Fossil Fuel Phase-out and What We Can Do about It*. London: Debt Justice.

World Bank. 2021a. *Climate Risk Country Profile: Argentina*. Washington, DC: World Bank.

World Bank. 2021b. *South African Economic Update, Edition 13: Building Back Better from COVID-19, with a Special Focus on Jobs*. Washington, DC: World Bank.

World Bank. 2022. *Senegal: First Equitable and Resilient Recovery Development Policy Financing*. Washington, DC: World Bank. World Bank Program Document.

World Bank. 2023a. *Colombia Country Climate and Development Report*. Washington, DC: World Bank.

World Bank. 2023b. "Delivery on the Ground: Country Action for a Livable Planet." https://live.worldbank.org/en/event/2023/2023-annual-meetings-delivery-on-the-ground-country-action-for-a-livable-planet (February 29, 2024).

World Bank and Asian Development Bank. 2021a. *Climate Risk Country Profile: Indonesia*. Washington, DC: World Bank & Asian Development Bank.

World Bank and Asian Development Bank. 2021b. *Climate Risk Country Profile: Pakistan*. Washington, DC: World Bank & Asian Development Bank.

World Resources Institute. 2024a. "Climate Watch: Argentina." www.climatewatchdata.org/countries/ARG (January 29, 2025).

World Resources Institute. 2024b. "Climate Watch: Colombia." www.climatewatchdata.org/countries/COL (December 18).

World Resources Institute. 2024c. "Climate Watch: Indonesia." www.climatewatchdata.org/countries/IDN (December 12).

World Resources Institute. 2024d. "Climate Watch: Pakistan." www.climatewatchdata.org/countries/PAK (January 29).

World Resources Institute. 2024e. "Climate Watch: South Africa." www.climatewatchdata.org/countries/ZAF (December 16).

Xie, Lina, Bert Scholtens, and Swarnodeep Homroy. 2023. "Rebalancing Climate Finance: Analysing Multilateral Development Banks' Allocation Practices." *Energy Research & Social Science* 101: 103127. https://doi.org/10.1016/j.erss.2023.103127.

Acknowledgments

Open access financing was generously provided through an ESRC New Investigator Grant (ES/V012916/1). We are also grateful to the Recourse Foundation – in particular, Federico Sibaja, Nezir Sinani, and Friederike Strub – for supporting our work.

Cambridge Elements

Organizational Response to Climate Change

Aseem Prakash
University of Washington

Aseem Prakash is Professor of Political Science, the Walker Family Professor for the College of Arts and Sciences, and the Founding Director of the Center for Environmental Politics at University of Washington, Seattle. His recent awards include the American Political Science Association's 2020 Elinor Ostrom Career Achievement Award in recognition of "lifetime contribution to the study of science, technology, and environmental politics," the International Studies Association's 2019 Distinguished International Political Economy Scholar Award that recognizes "outstanding senior scholars whose influence and path-breaking intellectual work will continue to impact the field for years to come," and the European Consortium for Political Research Standing Group on Regulatory Governance's 2018 Regulatory Studies Development Award that recognizes a senior scholar who has made notable "contributions to the field of regulatory governance."

Jennifer Hadden
University of Maryland

Jennifer Hadden is Associate Professor in the Department of Government and Politics at the University of Maryland. She conducts research in international relations, environmental politics, network analysis, nonstate actors, and social movements. Her research has been published in various journals, including the *British Journal of Political Science, International Studies Quarterly, Global Environmental Politics, Environmental Politics,* and *Mobilization*. Dr. Hadden's award-winning book, *Networks in Contention: The Divisive Politics of Global Climate Change,* was published by Cambridge University Press in 2015. Her research has been supported by a Fulbright Fellowship, as well as grants from the National Science Foundation, the National Socio-Environmental Synthesis Center, and others. She held an International Affairs Fellowship from the Council on Foreign Relations for the 2015–16 academic year, supporting work on the Paris Climate Conference in the Office of the Special Envoy for Climate Change at the US Department of State.

David Konisky
Indiana University

David Konisky is Professor at the Paul H. O'Neill School of Public and Environmental Affairs, Indiana University, Bloomington. His research focuses on US environmental and energy policy, with particular emphasis on regulation, federalism and state politics, public opinion, and environmental justice. His research has been published in various journals, including the *American Journal of Political Science, Climatic Change,* the *Journal of Politics, Nature Energy,* and *Public Opinion Quarterly*. He has authored or edited six books on environmental politics and policy, including *Fifty Years at the U.S. Environmental Protection Agency: Progress, Retrenchment and Opportunities* (Rowman & Littlefield, 2020, with Jim Barnes and John D. Graham), *Failed Promises: Evaluating the Federal Government's Response to Environmental Justice* (MIT, 2015), and *Cheap and Clean: How Americans Think about Energy in the Age of Global Warming* (MIT, 2014, with Steve Ansolabehere). Konisky's research has been funded by the National Science Foundation, the Russell Sage Foundation, and the Alfred P. Sloan Foundation. Konisky is currently coeditor of *Environmental Politics*.

Matthew Potoski
UC Santa Barbara

Matthew Potoski is a Professor at UCSB's Bren School of Environmental Science and Management. He currently teaches courses on corporate environmental management, and his research focuses on management, voluntary environmental programs, and public policy. His research has appeared in business journals such as *Strategic Management Journal, Business Strategy and the Environment,* and the *Journal of Cleaner Production,* as well as public policy and management journals such as *Public Administration Review* and the *Journal of Policy Analysis and Management.* He coauthored *The Voluntary Environmentalists* (Cambridge, 2006) and *Complex Contracting* (Cambridge, 2014; the winner of the 2014 Best Book Award, American Society for Public Administration, Section on Public Administration Research) and was coeditor of *Voluntary Programs* (MIT, 2009). Professor Potoski is currently coeditor of the *Journal of Policy Analysis and Management* and the *International Public Management Journal.*

About the Series

How are governments, businesses, and nonprofits responding to the climate challenge in terms of what they do, how they function, and how they govern themselves? This series seeks to understand why and how they make these choices and with what consequence for the organization and the eco-system within which it functions.

Cambridge Elements

Organizational Response to Climate Change

Elements in the Series

Explaining Transformative Change in ASEAN and EU Climate Policy: Multilevel Problems, Policies and Politics
Charanpal Bal, David Coen, Julia Kreienkamp, Paramitaningrum, and Tom Pegram

Fighting Climate Change through Shaming
Sharon Yadin

Governing Sea Level Rise in a Polycentric System: Easier Said than Done
Francesca Vantaggiato and Mark Lubell

Inside the IPCC: How Assessment Practices Shape Climate Knowledge
Jessica O'Reilly, Mark Vardy, Kari De Pryck, and Marcela da S. Feital Benedetti

Climate Activism, Digital Technologies, and Organizational Change
Mette Eilstrup-Sangiovanni and Nina Hall

Who Tells Your Story? Women and Indigenous Peoples Advocacy at the UNFCCC
Takumi Shibaike and Bi Zhao

Climate Adaptation and Conflict Mitigation: The Case of South Sudan
Ore Koren and Jerry Urtuzuastigui

Rebel Governance in the Age of Climate Change
Kathleen Gallagher Cunningham, Leonardo Gentil-Fernandes, Elisabeth Gilmore, Reyko Huang, Danielle F. Jung and Cyanne E. Loyle

Greening the International Monetary Fund
Alexandros Kentikelenis and Thomas Stubbs

A full series listing is available at: www.cambridge.org/ORCC

For EU product safety concerns, contact us at Calle de José Abascal, 56–1°, 28003 Madrid, Spain or eugpsr@cambridge.org.

www.ingramcontent.com/pod-product-compliance
Lightning Source LLC
LaVergne TN
LVHW011852060526

838200LV00054B/4293